ASCE

American Society
of Civil Engineers

# Civil Engineering Body of Knowledge for the 21st Century

## Preparing the Civil Engineer for the Future

Second Edition

Prepared by the
Body of Knowledge
Committee
of the
Committee on
Academic
Prerequisites for
Professional
Practice

American Society of Civil Engineers
1801 Alexander Bell Drive
Reston, Virginia, 20191-4400

www.asce.org


Manufactured in the United States of America.

**Library of Congress Cataloging-in-Publication Data**

Civil engineering body of knowledge for the 21st century: preparing the civil engineer for the future/prepared by the Body of Knowledge Committee of the Committee on Academic Prerequisites for Professional Practice.—2nd ed.
p. cm.
Includes bibliographical references and index.
ISBN-13: 978-0-7844-0965-7
ISBN-10: 0-7844-0965-X
1. Civil engineering—Study and teaching (Higher)—United States. 2. Civil engineering—Vocational guidance—United States. I. American Society of Civil Engineers. Body of Knowledge Committee.

T73.C538 2008
624.023—dc22

2007051364

# Contents

# List of Figures

# List of Tables

# List of Tables

# Executive Summary

*Destiny is not a matter of chance,*
*it is a matter of choice.*
William Jennings Bryan, American statesman

*The civil engineering profession is proactively preparing for the future.*

The manner in which civil engineering is practiced must change. That change is necessitated by such forces as globalization, sustainability requirements, emerging technology, and increased complexity with the corresponding need to identify, define, and solve problems at the boundaries of traditional disciplines. As always within the civil engineering profession, change must be accomplished mindful of the profession's primary concern for protecting public safety, health, and welfare.

The profession recognizes the need for change. For example, in June 2006, the American Society of Civil Engineers (ASCE) convened the Summit on the Future of Civil Engineering – 2025. This gathering of civil engineering and other leaders, including international participants, articulated a global vision for the future of civil engineering. The vision[1] sees civil engineers as being entrusted by society as leaders in creating a sustainable world and enhancing the global quality of life.

## Body of Knowledge

*ASCE's Policy Statement 465 calls for attainment of a body of knowledge for entry into the practice of civil engineering at the professional level.*

Even before the 2006 summit, the profession recognized the need for change. Beginning in 1998, ASCE's Board of Direction adopted, refined, and confirmed ASCE Policy Statement (PS) 465—"Academic Prerequisites for Licensure and Professional Practice"[2]—which "...supports the attainment of a body of knowledge (BOK) for entry into the practice of civil engineering at the professional level." The policy explains that this "...would be accomplished through the adoption of appropriate engineering education and experience requirements as a prerequisite for licensure." PS 465 recognizes that the profession's principal means of changing the way civil engineering is practiced lies in reforming the manner in which tomorrow's civil engineers are prepared—through education and early experience—to enter professional practice.

The permanent board-level Committee on Academic Prerequisites for Professional Practice ($CAP^3$) is charged with implementing PS 465. $CAP^3$ developed an implementation master plan, of which the BOK is the foundation. As one of its actions, $CAP^3$ created a BOK committee, which published the first BOK (BOK1) in January 2004. In response to the expanding use of BOK1 by various stakeholders, and the questions asked and suggestions offered as a result of that use, $CAP^3$ formed the second BOK Committee in October 2005. This committee was asked to produce a second edition of the BOK report in response to stakeholder input and recent developments in engineering education and practice. The result is the refined BOK (BOK2) presented in this report.

The BOK2 committee began its work by reviewing the 15 outcomes making up the core of BOK1.[3] Also examined were recent National Academy of Engineering reports,[4,5] which aligned with BOK1, and other documents. Outcomes are the heart of the BOK because they define the knowledge, skills, and attitudes necessary to enter the practice of civil engineering at the professional level in the 21st century.

*The original BOK was refined in response to stakeholder input and recent developments in engineering education and practice.*

Following careful deliberation, the original set of 15 outcomes was expanded to 24 outcomes organized into three categories: foundational, technical, and professional. The evolution from 15 to 24 outcomes further describes the BOK. Rather than add content, the larger number of outcomes add specificity and clarity. (See Appendix H for more detail.)

The committee adopted Bloom's Taxonomy, which is widely known and understood within the education community, as the means of describing the minimum cognitive levels of achievement for each outcome. Figure ES-1 presents the 24 outcomes and, for each one, the level of achievement that an individual should demonstrate to enter the practice of civil engineering at the professional level.

## Fulfilling the Body of Knowledge

*The BOK will be fulfilled by a combination of education and experience.*

According to PS 465, the BOK will be fulfilled by means of formal education and experience—that is, a bachelor's degree plus a master's degree, or approximately 30 semester

| Outcome Number and Title | Level of Achievement | | | | | |
|---|---|---|---|---|---|---|
| | 1 | 2 | 3 | 4 | 5 | 6 |
| | Knowledge | Compre-hension | Application | Analysis | Synthesis | Evaluation |
| *Foundational* | | | | | | |
| 1. Mathematics | B | B | B | | | |
| 2. Natural sciences | B | B | B | | | |
| 3. Humanities | B | B | B | | | |
| 4. Social sciences | B | B | B | | | |
| *Technical* | | | | | | |
| 5. Materials science | B | B | B | | | |
| 6. Mechanics | B | B | B | B | | |
| 7. Experiments | B | B | B | B | M/30 | |
| 8. Problem recognition and solving | B | B | B | M/30 | | |
| 9. Design | B | B | B | B | B | E |
| 10. Sustainability | B | B | B | E | | |
| 11. Contemp. issues & hist. perspectives | B | B | B | E | | |
| 12. Risk and uncertainty | B | B | B | E | | |
| 13. Project management | B | B | B | E | | |
| 14. Breadth in civil engineering areas | B | B | B | B | | |
| 15. Technical specialization | B | M/30 | M/30 | M/30 | M/30 | E |
| *Professional* | | | | | | |
| 16. Communication | B | B | B | B | E | |
| 17. Public policy | B | B | E | | | |
| 18. Business and public administration | B | B | E | | | |
| 19. Globalization | B | B | B | E | | |
| 20. Leadership | B | B | B | E | | |
| 21. Teamwork | B | B | B | E | | |
| 22. Attitudes | B | B | E | | | |
| 23. Lifelong learning | B | B | B | E | E | |
| 24. Professional and ethical responsibility | B | B | B | B | E | E |

Key:

| | |
|---|---|
| B | Portion of the BOK fulfilled through the bachelor's degree |
| M/30 | Portion of the BOK fulfilled through the master's degree or equivalent (approximately 30 semester credits of acceptable graduate-level or upper-level undergraduate courses in a specialized technical area and/or professional practice area related to civil engineering) |
| E | Portion of the BOK fulfilled through the prelicensure experience |

Figure ES-1. Entry into the practice of civil engineering at the professional level requires fulfilling 24 outcomes to the appropriate levels of achievement.

credits, and experience. Two common fulfillment paths were developed—one involving an accredited bachelor's degree in civil engineering followed by a master's degree, or approximately 30 semester credits of acceptable graduate-level or upper-level undergraduate courses, and the other using an appropriate bachelor's degree followed by an accredited master's degree.

The roles of the bachelor's degree, the master's degree or approximately 30 credits, and experience in fulfilling the BOK are shown in Figure ES-1. A detailed version of the figure, known as an outcome rubric, appears as Appendix I and non-prescriptive explanations for outcomes are presented in Appendix J. These two appendices are the heart of this report. The report presents two models for validating the fulfillment of the BOK, one for each of the two previously mentioned common fulfillment paths.

*The refined BOK is the foundation of the Policy Statement 465 Master Plan.*

This report stresses the foundational role of the BOK in implementing the PS 465 Master Plan, noting how the CAP$^3$ committee and its subcommittees build on BOK2. Also presented are ways the BOK could be used by prospective civil engineering students, high school counselors, parents, employers, and others.

## Roles of Faculty, Students, Engineer Interns, and Practitioners

*This report offers guidance to BOK stakeholders.*

PS 465 and the foundational BOK will reform the education and prelicensure experience of tomorrow's civil engineers. The resulting changes may raise concerns for some faculty members, students, engineer interns, and those practitioners who recruit, employ, supervise, coach, or mentor engineer interns. Accordingly, the BOK2 Committee invited various accomplished professionals, drawn from academia and practice and from the private and public sectors, to offer guidance ideas. Their input was used by the committee to create separate guidance for faculty, students, interns, and practitioners. That guidance is offered in this report in the hope that it provides useful insights and advice.

## The Next Steps

The BOK2 Committee believes that this report will significantly assist with further implementation of ASCE PS 465. Accordingly, the report concludes with implementation recommendations for many stakeholders, including the CAP[3] accreditation, licensure, educational fulfillment, and experience fulfillment committees; university departments of civil and environmental engineering; employers of civil engineers; civil engineering students and interns; and other engineering disciplines and organizations.

*The report concludes with recommendation for using the BOK to continue implementation of ASCE Policy Statement 465.*

# CHAPTER 1

# Introduction

*If you want to build a ship, don't drum up the people to gather wood, divide up the work, and give orders. Instead, teach them to yearn for the vast and endless sea.*
Antoine de Saint-Exupéry, French poet

## The Vision for Civil Engineering in 2025

*The vision: civil engineers will be entrusted by society to create a sustainable world and enhance the quality of life.*

The American Society of Civil Engineers defines civil engineering as "...the profession in which a knowledge of the mathematical and physical sciences gained by study, experience, and practice is applied with judgment to develop ways to utilize, economically, the materials and forces of nature for the progressive well-being of humanity in creating, improving and protecting the environment, in providing facilities for community living, industry and transportation, and in providing structures for the use of humanity."[6] The civil engineering profession is moving forward.

For example, in June 2006, a diverse group of civil engineering and other leaders, including international participants, gathered to formulate a global vision for the future of civil engineering. The term "vision," as applied at the summit, was conceptual as opposed to concrete; mental, not today's reality. It was reflective of actual or desired values, influenced by imagination, and focused on what, not how.[7]

Participants in this summit envisioned a very different world for civil engineers in 2025. An ever-increasing global population that is shifting even more to urban areas will require widespread adoption of sustainability. Demands for energy, transportation, drinking water, clean air, and safe waste disposal will drive environmental protection and infrastructure development. Society will face threats from natural events, accidents, and perhaps such other causes as terrorism.

Informed by the preceding, a global vision was developed that sees civil engineers entrusted by society to lead in creating a sustainable world and enhancing the global quality of life. The 2025 vision is:[1]

> *Entrusted by society to create a sustainable world and enhance the global quality of life, civil engineers serve competently, collaboratively, and ethically as master:*
>
> - *planners, designers, constructors, and operators of society's economic and social engine, the built environment;*
> - *stewards of the natural environment and its resources;*
> - *innovators and integrators of ideas and technology across the public, private, and academic sectors;*
> - *managers of risk and uncertainty caused by natural events, accidents, and other threats; and*
> - *leaders in discussions and decisions shaping public environmental and infrastructure policy.*

As used in the vision, "master" means one who possesses widely recognized and valued knowledge, skills, and attitudes acquired as a result of education, experience, and achievement. Individuals within a profession who have these characteristics are willing and able to serve society by orchestrating solutions to society's most pressing current needs while helping to create a more viable future.

Summit organizers and participants intend that the vision will guide policies, plans, processes, and progress within the civil engineering community and beyond, worldwide. Civil engineers and leaders of civil engineering organizations should act to move the civil engineering profession toward the vision. One critical action is reform in the education and prelicensure experience of civil engineers. This report is presented in the spirit of that reform.

*Reform in education and prelicensure experience will help achieve the civil engineering vision.*

## ASCE Policy Statement 465: Emergence of the Body of Knowledge

*The ASCE Board of Direction adopted Policy Statement 465 in 1998 and thus began reformation of the education and prelicensure experience of civil engineers.*

*ASCE supports the attainment of a BOK for entry into the practice of civil engineering at the professional level.*

In October 1998, following years of studies and conferences, the ASCE Board of Direction adopted Policy Statement 465 (PS 465), which has since been refined and confirmed. The ASCE board last revised Policy Statement 465 in April 2007 and, as it has since October 2004, the statement explicitly includes the body of knowledge (BOK).[8] The policy now reads, in part:

> *The ASCE supports the attainment of a body of knowledge for entry into the practice of civil engineering at the professional level. This would be accomplished through the adoption of appropriate engineering education and experience requirements as a prerequisite for licensure.*

The BOK is defined in the policy as "the necessary depth and breadth of knowledge, skills, and attitudes required of an individual entering the practice of civil engineering at the professional level in the 21st century." The long-term effect of PS 465 is illustrated in Figure 1, which compares today's civil engineering professional track with tomorrow's.

From ASCE's perspective, the civil engineering BOK represents a strategic direction for the profession. Under today's curricula design and accreditation and regulatory processes and procedures, some of the elements of the BOK may not be translated into curricula, accreditation criteria, and licensing requirements in the near term. In other words, the BOK describes the "gold standard" for the aspiring civil engineering professional. Because input into curricula design, accreditation, and licensing comes from many and varied stakeholders beyond ASCE, these processes are not likely to reflect all aspects of the civil engineering BOK. ASCE is optimistic that the curricula design, accreditation, and licensing processes will change over time to adopt a more BOK-centric approach. As this occurs, a greater proportion of the BOK will be reflected in curricula and in accreditation and licensure requirements.

The BOK thrust resulted in the BOK Committee of the ASCE Committee on Academic Prerequisites for Professional Practice (CAP$^3$) completing, in January 2004, the report *Civil Engineering Body of Knowledge for the 21st Century: Preparing the Civil Engineer for the Future.*[3] Report recommendations were cast in terms of 15 outcomes that, compared to today's bachelor's programs, provided signifi-

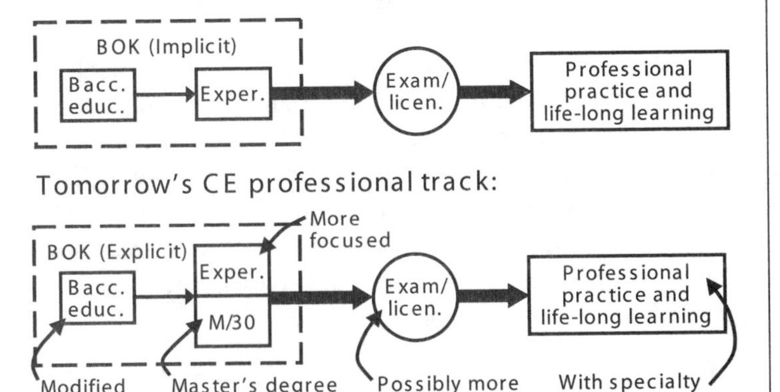


Figure 1. Implementation of Policy Statement 465 will improve the lifelong career of tomorrow's civil engineer.

cant increases in technical depth and professional practice breadth. Included in the 15 outcomes were the 11 outcomes directly influenced by those used by ABET in its General Criteria for Baccalaureate Level Programs. Each outcome was further described with a helpful, non-prescriptive civil engineering commentary.

As a result of reviewing and using the recommendations in the civil engineering BOK report, stakeholders identified a problem and raised issues related to the BOK. The problem was ambiguity of the three principal words used to define competency levels—namely, recognition, understanding, and ability.

To remove this obstacle, CAP[3] formed the Levels of Achievement Subcommittee in February 2005. The subcommittee's September 2005 report[9] contained many recommendations that addressed the problem and are being implemented. Relative to this second edition BOK report, the subcommittee recommended using Bloom's Taxonomy to define levels of achievement. Bloom's Taxonomy is explained in Chapter 2 of this second edition BOK report and applied in Chapter 2 and Chapter 3. Bloom's levels of the cognitive domain are widely known and understood within the education community. Furthermore, use of measurable, action-oriented verbs will facilitate more consistent curricula design and assessment. (Refer to Appendix C for a more detailed account of PS 465.)

## Formation of and Charge to the Second Body of Knowledge Committee

*The second BOK committee was formed to respond to questions and suggestions raised by the report produced by the first committee.*

In response to the recommendations of the Levels of Achievement Subcommittee, the expanding use of the civil engineering BOK by various stakeholders, and the questions asked and suggestions offered as a result of that use, $CAP^3$ formed the second Body of Knowledge committee (the BOK2 Committee) in October 2005. This action was anticipated because the initial BOK report was envisioned as a first edition—that is, a work in progress.

The charge to the BOK2 Committee appears as Appendix D. The public solicitation process used to select committee members, along with the names, affiliations, and contact information for committee members, are presented in Appendix E. $CAP^3$ asked the committee to produce a second edition of the BOK report in response to recent stakeholder input and other developments in engineering education and practice.

## Committee's Overall Approach

The BOK2 Committee carried out its work in an inclusive and transparent manner and adopted a proactive approach. It conducted 65 conference calls for the entire committee and held four face-to-face meetings during the period of October 2005 through November 2007, when a draft was completed for $CAP^3$ review.

*Corresponding members contributed significantly to the substance of this report.*

As soon as it was formed, the committee created a correspondents group to review draft materials, respond to questions, and otherwise provide ideas and information for consideration by the committee. Corresponding members are mostly civil engineers from the public and private sectors and academia. (Corresponding members and other contributors are listed in Appendix E.) Corresponding members, some of whom reside in other countries, participated in many e-mail discussions and frequently commented on draft materials. They contributed significantly to this report.

*Using a variety of mechanisms, committee members proactively interacted with stakeholders.*

During the course of their work, BOK2 Committee members spoke at and/or participated in various conferences, workshops and meetings, and wrote BOK-focused articles and papers for a wide variety of publications. These activities provided additional means of sharing progress and soliciting input.

# Note to the Reader

To assist the reader a list of abbreviations is included as Appendix A; Appendix B is a glossary. Many other appendices, providing various types of detailed information, appear at the end of the report and are cited at least once in the text. Some aspects of the committee's work required in-depth research. Accordingly, the committee documented that research in appendices for readers who may benefit from and build on it. Examples are Appendices F and G and K through O, which address Bloom's Taxonomy, the affective domain, the humanities and social sciences, sustainability, globalization, public policy, and attitudes. The last appendix, Appendix P—Notes—is keyed to the body of the report via superscript numbers.

# CHAPTER 2

# Body of Knowledge—Knowledge, Skills, and Attitudes Necessary for Entry into Professional Practice

*Engineers must be society wise*
*as well as technology wise.*
Warren Viessman, Jr., P.E., Hon.M.ASCE and civil engineer

## Introduction

*Outcomes define the knowledge, skills, and attitudes needed to enter the practice of civil engineering at the professional level in the 21st century.*

Refining the civil engineering BOK for the 21st century challenged the BOK2 Committee, just as defining the initial BOK challenged the first BOK Committee. The committee began the refinement process by reviewing the 15 outcomes presented in the original BOK report,[3] recent NAE reports,[4,5] and other relevant documents. For purposes of the civil engineering BOK, outcomes are statements that describe what individuals are expected to know and be able to do by the time of entry into the practice of civil engineering at the professional level in the 21st century—that is, attain licensure. Outcomes define the knowledge, skills, and attitudes that individuals acquire through appropriate formal education and prelicensure experience.

The committee focused on outcomes without consideration of courses, semesters, faculty expectations, co-curricular and extracurricular activities, access and delivery systems, and other administrative and logistical aspects of teaching and learning the outcomes. For example, topics listed in the outcomes could appear in more than one course, one course could contain many of the outcomes, and, conceivably, one outcome could encompass an entire course. Many outcomes will be partially fulfilled during prelicensure experience.

Task groups, composed of members of the BOK2 Committee and others, were formed to review the original 15 outcomes, to evaluate the need for revised and new outcomes, and to consider the possibility of consolidating outcomes. The idea was to determine if the original outcomes were still appropriate—that is, if they had stood the test of time over the several years that they have been available for discussion and use. This process, which was guided by Bloom's Taxonomy (described in the next section), led the committee to a refined set of outcomes, an explanation for each, and a BOK rubric. Civil engineering department heads reviewed the outcomes and provided detailed comments, which were used by the committee to further refine the outcomes, their explanations, and the rubric.

*The committee started with and thoroughly reviewed the original 15 outcomes.*

## Bloom's Taxonomy

Articulation of BOK outcomes and related levels of achievement reflects the desire to clarify what should be taught and learned. Clarification can be achieved through the use of organizing frameworks or taxonomies that systematically differentiate outcome characteristics and promote common understandings for all potential users of the BOK.

*Learning taxonomies help to articulate BOK outcomes and achievement levels.*

Accordingly, the ASCE Levels of Achievement Subcommittee,[9] which completed its work in September 2005, undertook a review of the educational psychology literature to find potential frameworks that might be applicable to the BOK. Specifically, the subcommittee wanted a relatively simple framework, informed by educational research, which could link BOK outcomes to actual learning and achievement. The taxonomy that met simplicity and relevancy needs was Bloom's Taxonomy, as discussed in more detail in Appendices F and G. Bloom's Taxonomy—published in the *Taxonomy of Educational Objectives, the Classification of Educational Goals, Handbook I: Cognitive Domain*[10]—continues to find use today. The handbook was "ahead of its time"[11] and its impact nationally and internationally is well documented.

*Bloom's Taxonomy satisfied simplicity and relevancy needs.*

In summary, Bloom's Taxonomy provides an appropriate framework for the definition of levels of achievement in the civil engineering BOK because:

- Bloom's Taxonomy is widely known and understood within the education community and its application to engineering education is documented in the literature. Thus levels of achievement based on Bloom's Taxonomy have broader legitimacy than any internally developed taxonomy would likely have.
- Bloom's emphasis on the use of measurable, action-oriented verbs linked to levels of development creates understandable and implementable outcome statements that will support consistent and more effective assessment.

Bloom's Taxonomy employs three distinct domains: the cognitive, the affective, and the psychomotor, which are described[10] as follows:

- "the cognitive domain ... includes those objectives [that] deal with the recall or recognition of knowledge and the development of intellectual abilities and skills."
- "the affective domain ... includes objectives [that] describe changes in interest, attitudes, and values ..."
- the psychomotor domain includes "the manipulative or motor-skill area."

This chapter focuses on the cognitive domain because that domain addresses many conventional learning outcomes associated with engineering. The affective domain is discussed in Appendix G. It has historically received less attention in engineering, although the BOK2 Committee considers its relevance noteworthy and potentially a useful complement to the cognitive domain in engineering. The third domain, psychomotor, is not pursued here.

## Outcomes: Introduction

*The 24 outcomes will broaden and deepen the formal education and prelicensure experience of civil engineers.*

Table 1 introduces the 24 outcomes—4 foundational outcomes, 11 technical outcomes, and 9 professional outcomes—recommended by the BOK2 Committee. Figure 2 summarizes the information graphically. The outcomes are organized by these three categories to further clarify the BOK.

The foundational category warrants some explanation. The four outcomes are foundational in two ways. First, these outcomes help to lay the foundation for the remaining technical and professional outcomes. They will help the civil engineering student or engineer intern fulfill the technical and professional outcomes. Second, these four outcomes are the foundation needed by a well-educated individual in the 21st century. Equipped with an understanding of mathematics, the natural sciences, and the humanities and social sciences, individuals will be in a position to understand the workings of the physical world and the behaviors of its inhabitants. The breadth of knowledge included in the foundational outcomes will also offer career and other life options.

The outcomes collectively describe the BOK—that is, the necessary depth and breadth of knowledge, skills, and attitudes required of an individual aspiring to enter the practice of civil engineering at the professional level in the 21st century. Relative to today's approach, tomorrow's civil engineer—prior to entry into the practice of civil engineering at the professional level—will:

*Relative to today's approach, the 24 outcomes in the refined BOK add technical depth and professional practice breadth.*

- master more mathematics, natural sciences, and engineering science fundamentals;
- maintain technical breadth;
- acquire broader exposure to the humanities and social sciences;
- gain additional professional practice breadth; and
- achieve greater technical depth—that is, specialization.

In Table 1 and Figure 2 outcomes are listed in approximate pedagogical order within the foundational, technical, and professional categories. Other than approximate pedagogical order, no inference, such as relative importance, should be drawn from the order in which outcomes are listed. Also, outcomes are not necessarily of equal importance even when they share identical recommended levels of achievement. Note how active verbs, indicated in bold italics in Table 1 and consistent with Bloom's Taxonomy, help define the level of achievement recommended for entry into the practice of civil engineering at the professional level.

*Active verbs define, for each outcome, the necessary level of achievement.*

## Outcomes: Enhanced Clarity

The evolution from 15 outcomes in the BOK1 report to 24 outcomes in this BOK2 report warrants discussion. In reviewing the original 15 outcomes, the committee determined that

Table 1. Entry into the practice of civil engineering at the professional level requires fulfilling 24 outcomes to the various levels of achievement.

**Key: L1** through **L6** refers to these levels of achievement:
Level 1 (**L1**) - Knowledge
Level 2 (**L2**) - Comprehension
Level 3 (**L3**) - Application
Level 4 (**L4**) - Analysis
Level 5 (**L5**) - Synthesis
Level 6 (**L6**) - Evaluation

| Outcome number and title | To enter the practice of civil engineering at the professional level, an individual must be able to demonstrate this level of achievement. |
|---|---|
| *Foundational Outcomes* | |
| 1<br>Mathematics | ***Solve*** problems in mathematics through differential equations and ***apply*** this knowledge to the solution of engineering problems. **(L3)** |
| 2<br>Natural sciences | ***Solve*** problems in calculus-based physics, chemistry, and one additional area of natural science and ***apply*** this knowledge to the solution of engineering problems. **(L3)** |
| 3<br>Humanities | ***Demonstrate*** the importance of the humanities in the professional practice of engineering **(L3)** |
| 4<br>Social sciences | ***Demonstrate*** the incorporation of social sciences knowledge into the professional practice of engineering. **(L3)** |
| *Technical Outcomes* | |
| 5<br>Materials science | **Use** knowledge of materials science to ***solve*** problems appropriate to civil engineering. **(L3)** |
| 6<br>Mechanics | **Analyze** and solve problems in solid and fluid mechanics. **(L4)** |
| 7<br>Experiments | ***Specify*** an experiment to meet a need, conduct the experiment, and analyze and ***explain*** the resulting data. **(L5)** |
| 8<br>Problem recognition and solving | ***Formulate*** and solve an ill-defined engineering problem appropriate to civil engineering by ***selecting*** and applying appropriate techniques and tools. **(L4)** |
| 9<br>Design | ***Evaluate*** the design of a complex system, component, or process and ***assess*** compliance with customary standards of practice, user's and project's needs, and relevant constraints. **(L6)** |
| 10<br>Sustainability | ***Analyze*** systems of engineered works, whether traditional or emergent, for sustainable performance. **(L4)** |
| 11<br>Contemporary issues and historical perspectives | ***Analyze*** the impact of historical and contemporary issues on the identification, formulation, and solution of engineering problems and ***analyze*** the impact of engineering solutions on the economy, environment, political landscape, and society. **(L4)** |

Table 1. Entry into the practice of civil engineering at the professional level requires fulfilling 24 outcomes to the various levels of achievement. (Continued)

| Outcome | Description |
|---|---|
| 12<br>Risk and uncertainty | ***Analyze*** the loading and capacity, and the effects of their respective uncertainties, for a well-defined design and ***illustrate*** the underlying probability of failure (or nonperformance) for a specified failure mode. **(L4)** |
| 13<br>Project management | ***Formulate*** documents to be incorporated into the project plan. **(L4)** |
| 14<br>Breadth in civil engineering areas | ***Analyze*** and solve well-defined engineering problems in at least four technical areas appropriate to civil engineering. **(L4)** |
| 15<br>Technical specialization | ***Evaluate*** the design of a complex system or process, or ***evaluate*** the validity of newly created knowledge or technologies in a traditional or emerging advanced specialized technical area appropriate to civil engineering. **(L6)** |
| *Professional Outcomes* | |
| 16<br>Communication | ***Plan, compose,*** and ***integrate*** the verbal, written, virtual, and graphical communication of a project to technical and non-technical audiences. **(L5)** |
| 17<br>Public policy | ***Apply*** public policy process techniques to simple public policy problems related to civil engineering works. **(L3)** |
| 18<br>Business and public administration | ***Apply*** business and public administration concepts and processes. **(L3)** |
| 19<br>Globalization | ***Analyze*** engineering works and services in order to function at a basic level in a global context. **(L4)** |
| 20<br>Leadership | ***Organize*** and ***direct*** the efforts of a group. **(L4)** |
| 21<br>Teamwork | ***Function*** effectively as a member of a multidisciplinary team. **(L4)** |
| 22<br>Attitudes | ***Demonstrate*** attitudes supportive of the professional practice of civil engineering. **(L3)** |
| 23<br>Lifelong learning | ***Plan*** and ***execute*** the acquisition of required expertise appropriate for professional practice. **(L5)** |
| 24<br>Professional and ethical responsibility | ***Justify*** a solution to an engineering problem based on professional and ethical standards and ***assess*** personal professional and ethical development. **(L6)** |

some of the individual outcomes could be clarified if they were presented as two or more outcomes.

For example, BOK1 outcome 1—the technical core—nominally becomes, in this BOK2 report, four outcomes: 1 (mathematics), 2 (natural sciences), 5 (materials science), and 6 (mechanics). Similarly, BOK1 outcome 10 (contemporary issues) nominally becomes outcome 11 (contemporary issues and historical perspectives) and outcome 19 (globalization). (These and the other relationships between the 15 BOK1 outcomes and the 24 BOK2 outcomes are shown in Appendix H.)

While some outcomes were added to further clarify the BOK1 outcomes, other outcomes were added to clarify other aspects of the BOK. For example, consider ABET's Criteria for Accrediting Engineering Programs.[12] "Criterion 4, Professional Component" in the General Criteria for Basic Level Programs states that one required element of the professional component is "a general education component that complements the technical content of the curriculum and is consistent with the program and institution objectives." Given the importance of this general education requirement and the effort expended on general education in the typical civil engineering curriculum, the BOK2 Committee endorsed this important requirement and concluded that it should be defined in terms of outcomes. Accordingly, the committee added outcome 3 (humanities) and outcome 4 (social sciences).

The ABET Program Criteria for Civil and Similarly Named Engineering Programs[12] calls for "proficiency in a minimum of four recognized major civil engineering areas." The BOK2 Committee concluded that this well-intentioned requirement should be clarified with the assistance of Bloom's Taxonomy. The result is outcome 14, breadth in civil engineering areas.

*The larger number of outcomes provides specificity and clarity without adding to the time required to fulfill them.*

In summary, the evolution from 15 outcomes to 24 outcomes further describes the BOK. The larger number of outcomes provides specificity and clarity, without adding to the time required to fulfill the BOK through formal education and prelicensure experience.

Reaching the recommended levels of achievement to fulfill the BOK will be accomplished through a combination of formal education and early experience. The next major section of this report presents a comprehensive discussion of fulfilling the BOK.

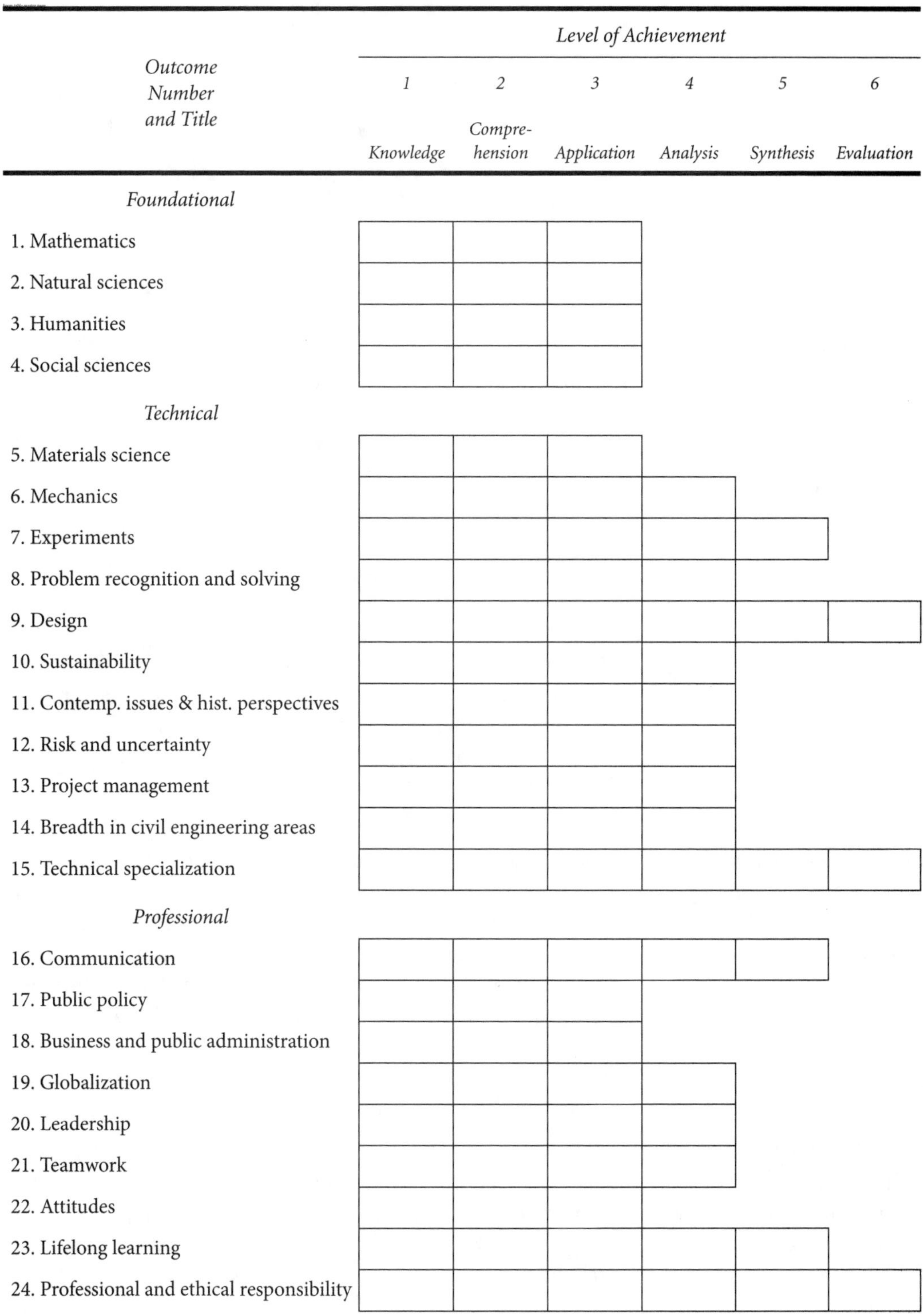

| Outcome Number and Title | Level of Achievement | | | | | |
|---|---|---|---|---|---|---|
| | 1 | 2 | 3 | 4 | 5 | 6 |
| | *Knowledge* | *Compre-hension* | *Application* | *Analysis* | *Synthesis* | *Evaluation* |
| *Foundational* | | | | | | |
| 1. Mathematics | | | | | | |
| 2. Natural sciences | | | | | | |
| 3. Humanities | | | | | | |
| 4. Social sciences | | | | | | |
| *Technical* | | | | | | |
| 5. Materials science | | | | | | |
| 6. Mechanics | | | | | | |
| 7. Experiments | | | | | | |
| 8. Problem recognition and solving | | | | | | |
| 9. Design | | | | | | |
| 10. Sustainability | | | | | | |
| 11. Contemp. issues & hist. perspectives | | | | | | |
| 12. Risk and uncertainty | | | | | | |
| 13. Project management | | | | | | |
| 14. Breadth in civil engineering areas | | | | | | |
| 15. Technical specialization | | | | | | |
| *Professional* | | | | | | |
| 16. Communication | | | | | | |
| 17. Public policy | | | | | | |
| 18. Business and public administration | | | | | | |
| 19. Globalization | | | | | | |
| 20. Leadership | | | | | | |
| 21. Teamwork | | | | | | |
| 22. Attitudes | | | | | | |
| 23. Lifelong learning | | | | | | |
| 24. Professional and ethical responsibility | | | | | | |

Figure 2. Entry into the practice of civil engineering at the professional level requires fulfilling 24 outcomes to the appropriate levels of achievement.

# CHAPTER 3

# Fulfilling the Body of Knowledge

*Learning is a treasure that*
*will follow its owner everywhere.*
Chinese proverb

## Introduction

The preceding chapter of this report introduced the revised BOK necessary for entry into the professional practice of civil engineering. The content of the preceding section might be considered the "what," as in what knowledge, skills, and attitudes are required to enter professional practice? Building on that introduction, this chapter focuses on the "how" and addresses this in two ways.

*This chapter discusses how tomorrow's aspiring civil engineer could fulfill the BOK.*

The first way is "how" the BOK could be fulfilled by tomorrow's aspiring civil engineer, using formal education and prelicensure experience,[13] to address the 24 foundational, technical, and professional outcomes at the designated levels of achievement. More specifically, this chapter describes two paths to fulfillment, presents a rubric that indicates the roles of education and prelicensure experience, introduces explanations for each of the outcomes, and outlines options for validating BOK fulfillment.

*The BOK is the foundation of the master plan for implementing ASCE Policy Statement 465.*

The second way this chapter addresses "how" is by explaining "how" the BOK is the foundation of the master plan for implementing ASCE PS 465. Related to this is a presentation of "how" the BOK could be used by various civil engineering stakeholders.

## Outcomes: Paths to Fulfillment

ASCE PS 465 states that fulfillment of the BOK will include a combination of:

- a baccalaureate degree in civil engineering;
- a master's degree, or approximately 30 coordinated graduate or upper-level undergraduate semester credits or the equivalent agency/organization/professional society courses providing equal quality and rigor; and
- appropriate experience based upon broad technical and professional practice guidelines that provide sufficient flexibility for a wide range of roles in engineering practice.

*BOK fulfillment will require a baccalaureate degree, a master's degree or equivalent, and experience.*

In symbolic form, this portion of PS 465 is referred to as:

**B + M/30 & E**

This means the "bachelor's plus master's, or approximately 30 semester credits of acceptable graduate-level or upper-level undergraduate courses in a specialized technical area and/or professional practice area related to civil engineering, and experience." Briefly, this may be stated as bachelor's plus either master's or approximately 30 credits, and experience. The B + M/30 portion of this expression represents several different but related methods to fulfill the formal educational component of the BOK. The E refers to progressive, structured engineering experience that, when combined with the educational requirements, results in attainment of the requisite civil engineering BOK. Two common fulfillment paths are described in the next two sections of this chapter.

To understand the discussions of the BOK fulfillment and validation that follow and as noted in Chapter 1, understand that some of the elements of the BOK may not be translated into accreditation criteria and experience requirements in the near term. Because input into the accreditation and licensing processes comes from many stakeholders beyond ASCE, these processes are not likely to reflect all aspects of the civil engineering BOK. Therefore, the two paths described below do not guarantee the total fulfillment and absolute validation of the BOK. ASCE is optimistic that the accreditation and licensing processes will change over time to adopt a more BOK-centric approach. As this occurs a growing proportion of the BOK will be reflected in the accreditation and licensure requirements.

*Path 1 begins with an ABET-accredited baccalaureate degree.*

### *Path 1 to Fulfillment*

Path 1 for fulfilling the BOK in the future can be symbolized as:

$$\mathbf{B^{ABET} \quad + \quad (M/30)^{Validated} \quad \& \quad E}$$

The B refers to a baccalaureate degree in civil engineering accredited by the Engineering Accreditation Commission of ABET, Inc. (EAC/ABET). The M/30 refers to a master's degree or equivalent (approximately 30 semester credits of acceptable graduate-level or upper-level undergraduate courses in a specialized technical area and/or professional practice area related to civil engineering). The M signifies a program leading to a master's degree that is not necessarily accredited by EAC/ABET. The 30-credit approach does not have to lead to a master's degree. The M/30 must be validated. Possible ways to do this are discussed later in this chapter in the "Outcomes: Validating Fulfillment" section.

The M or the 30 portion of this path can be accomplished equally well by traditional campus-based courses or by distance learning delivery systems, or a combination of the two. In the future, all of the 30 might be delivered through independently evaluated, high-quality, standards-based educational programs offered by firms, government agencies, professional societies, and for-profit educational organizations. Clearly, distance learning and independent educational programs are likely to become more prevalent and important in the future for both degree and non-degree granting programs.

### *Path 2 to Fulfillment*

Path 2 for fulfilling the BOK in the future can be symbolized as:

$$\mathbf{B \quad + \quad M^{ABET} \quad \& \quad E}$$

*Path 2 includes an ABET-accredited master's degree.*

While the baccalaureate degree associated with this path is not required to be an ABET/EAC accredited degree in civil engineering, the master's degree is an ABET/EAC accredited degree in or related to civil engineering. ASCE has pursued important modifications to ABET accreditation criteria and policies to make this a viable alternative path in the future.

In addition to paths 1 and 2 as explained above, ASCE continues to explore other paths for fulfilling the civil engineering BOK. Some of the additional paths being explored are for those who have a bachelor's degree in other engineering disciplines, a bachelor's degree not in engineering, or a bachelor's degree in engineering technology, but who are unable to pursue path 2 described above. ASCE wishes to attract individuals to the civil engineering profession from nontraditional routes—while trying to ensure that all of those who enter the professional practice of civil engineering have the necessary knowledge, skills, and attitudes.

## Outcomes: Rubric

*The outcome rubric presents the outcomes, required levels of achievement for each, and the roles of education and experience.*

Building on the recommendations of the Levels of Achievement Subcommittee,[19] the BOK2 Committee developed the outcome rubric[14] presented in detail as Appendix I and summarized graphically in Figure 3. The rubric communicates the following BOK characteristics:

1. The 24 outcomes, categorized as foundational, technical, and professional and, within each category, organized in approximate pedagogical order while not reflecting relative importance.
2. The recommended level of achievement that an individual must demonstrate for each outcome to enter the practice of civil engineering at the professional level.
3. For each outcome, the portion of the level of achievement to be fulfilled through the bachelor's degree, the portion to be fulfilled through the master's degree or equivalent, and the portion to be fulfilled through prelicensure experience.

BOK features illustrated in Figure 3 and Appendix I include:

*The bachelor's degree is the foundation of all outcomes.*

1. All 24 outcomes—with the exception of outcome 15 (technical specialization)—are fulfilled, at least through level 2 (comprehension), via formal education in a baccalaureate program. The bachelor's degree lays the foundation for all outcomes and provides a broad background in the natural sciences, the humanities, the social sciences, and engineering.
2. For six outcomes (1, 2, 3, 4, 5, and 6), the necessary levels of achievement are fulfilled entirely through the bachelor's degree. Coupled with the preceding observation, this emphasizes the importance of a broad baccalaureate education that provides a solid foundation for higher-level education (M/30) and prelicensure experience.

| Outcome Number and Title | Level of Achievement | | | | | |
|---|---|---|---|---|---|---|
| | 1 | 2 | 3 | 4 | 5 | 6 |
| | *Knowledge* | *Compre-hension* | *Application* | *Analysis* | *Synthesis* | *Evaluation* |
| *Foundational* | | | | | | |
| 1. Mathematics | B | B | B | | | |
| 2. Natural sciences | B | B | B | | | |
| 3. Humanities | B | B | B | | | |
| 4. Social sciences | B | B | B | | | |
| *Technical* | | | | | | |
| 5. Materials science | B | B | B | | | |
| 6. Mechanics | B | B | B | B | | |
| 7. Experiments | B | B | B | B | M/30 | |
| 8. Problem recognition and solving | B | B | B | M/30 | | |
| 9. Design | B | B | B | B | B | E |
| 10. Sustainability | B | B | B | E | | |
| 11. Contemp. issues & hist. perspectives | B | B | B | E | | |
| 12. Risk and uncertainty | B | B | B | E | | |
| 13. Project management | B | B | B | E | | |
| 14. Breadth in civil engineering areas | B | B | B | B | | |
| 15. Technical specialization | B | M/30 | M/30 | M/30 | M/30 | E |
| *Professional* | | | | | | |
| 16. Communication | B | B | B | B | E | |
| 17. Public policy | B | B | E | | | |
| 18. Business and public administration | B | B | E | | | |
| 19. Globalization | B | B | B | E | | |
| 20. Leadership | B | B | B | E | | |
| 21. Teamwork | B | B | B | E | | |
| 22. Attitudes | B | B | E | | | |
| 23. Lifelong learning | B | B | B | E | E | |
| 24. Professional and ethical responsibility | B | B | B | B | E | E |

Key:

| | |
|---|---|
| B | Portion of the BOK fulfilled through the bachelor's degree |
| M/30 | Portion of the BOK fulfilled through the master's degree or equivalent (approximately 30 semester credits of acceptable graduate-level or upper-level undergraduate courses in a specialized technical area and/or professional practice area related to civil engineering) |
| E | Portion of the BOK fulfilled through the prelicensure experience |

Figure 3. The BOK rubric integrates outcomes, levels of achievement, formal education, and prelicensure experience.

3. The M/30 helps to fulfill three outcomes (7, 8, and 15) and is the primary means by which outcome 15, technical specialization, is accomplished. Outcome 15 and the supporting role of outcomes 7 and 8 at the M/30 level provide the greater technical depth in the BOK.
4. For 15 outcomes (outcomes 9 through 13 and 15 through 24), almost two thirds of the total, experience is needed, in addition to formal education, to enter the practice of civil engineering at the professional level. This reinforces the need for education/experience partnerships in fulfilling the BOK.
5. As suggested by the dominance of the B cells in Figure 3, most of the formal education in the BOK occurs during the bachelor's degree program. This provides an educational foundation shared across civil engineering. Accordingly, completion of a well-rounded and broad baccalaureate program at one institution offers the option of a transportable progression to a more focused and specialized master's degree or equivalent at another school or in another program.

*Experience is essential for fulfilling almost two thirds of the outcomes.*

Consider the portion of the rubric to the right of the cells containing the B, M/30, and E notations. The BOK2 Committee used a reverse process in developing the rubric. The committee first "filled in" the entire rubric—that is, all six of Bloom's Taxonomy levels for 24 outcomes (144 cells)—prior to selecting the levels of achievement needed for entry into the practice of civil engineering at the professional level. And only after those levels were established did the committee, working in reverse, make decisions concerning the recommended roles of B, M/30, and E.

After completing this systematic reverse process, the committee decided to retain the Bloom's Taxonomy-based information in the cells to the right of the B-M/30-E cells in Appendix I because some of that information may be useful to various rubric users. Three examples:

*The committee first established achievement levels needed for entry into professional practice and then addressed the roles of education and experience.*

- A faculty member, student, or engineer intern might note, for any outcome, the level of achievement in the cell immediately to the right of the cell defining the level of achievement needed at the completion of the B or M/30 or for entry into professional practice. The former will help the user further understand the latter.
- Another example of the potential usefulness of the "right side" of the rubric is based on the realization that the rubric defines the minimum level of achievement for each

outcome. An individual—a student or engineer intern, for example—or an entity, such as a civil engineering department or an engineering employer, may want to go beyond the suggested minimums. The higher levels of achievement described in the "right side" cells would be useful in defining going beyond minimums.

- Some ASCE institutes are engaged in specialty certification for which licensure is one requirement. Specialty certification criteria could build upon the BOK2 outcomes and their minimum achievement levels.

The process used by the committee to eventually establish the B, M/30, and E cells in the rubric could also be applied by educators and by engineering organization managers. For example, faculty designing new or revised undergraduate curricula, could, for each outcome, begin with the highest level of achievement for that outcome, within the baccalaureate program. Then, working in reverse, they could design or redesign related curricular elements. Similarly, leaders within engineering organizations could, for each outcome requiring experience, note the highest level of achievement for that outcome to be accomplished via experience. Then, working in reverse, they could design a coaching, mentoring, and education and training program for interns to assist them in fulfilling the BOK.

## Outcomes: Explanations

*Non-prescriptive explanations are provided to aid BOK stakeholders—that is, faculty, students, engineer interns, and practitioners.*

The BOK2 Committee developed explanations for each of the 24 outcomes. These explanations are designed to help faculty who teach aspiring civil engineers and practitioners who recruit, employ, supervise, coach, or mentor prelicensure civil engineers. The explanations will also aid both civil engineering students and civil engineer interns—that is, individuals who are preparing for entry into the professional practice of civil engineering. To reiterate, explanations are to be helpful in communicating the intent and content of outcomes—they are not prescriptive. Outcomes paired with explanations provide a desirable deliverable for stakeholders; Bloom's Taxonomy-based outcomes relying on active verbs, each outcome supported by a descriptive and illustrative explanation.

Outcomes are viewed as being applicable over a long period of time—years, for example. In contrast, some illustrative topics mentioned in the explanations will be ephemeral, requiring modification in response to stakeholder needs, technological advances, and other changes.

Using a format of one or two explanations per page, the explanations are presented in Appendix J. This format enables the reader to readily move from one outcome to another. The format for each explanation begins with an overview section that presents the rationale for the outcome and defines terms, as needed.

The overview is followed by a section, denoted by B, that states the minimum level of achievement to be fulfilled through the bachelor's degree. The level of achievement is taken directly from the rubric. The code L1, L2, L3, L4, L5, or L6 is included to reiterate, respectively, the following Bloom's Taxonomy level of achievement that is to be accomplished: knowledge, comprehension, application, analysis, synthesis, and evaluation. The B section goes on to offer ideas on curricular and, in some cases, co-curricular and extracurricular ways to enable the aspiring civil engineer to reach the required levels of achievement.

As appropriate for the outcome, the B section is followed by an M/30 (master's degree or equivalent) and/or an E (experience) section. As with the B section, these sections offer ideas on how an individual, within his or her courses or during his or her prelicensure experience, can attain the necessary minimum levels of achievement.

## Outcomes: Validating Fulfillment

Earlier in this chapter, two paths were presented for the fulfillment of the civil engineering BOK. These paths represent different but related fulfillment models. The two fulfillment paths also correspond to two models for validating an individual's attainment of the BOK.

### *Validation of Path 1 to Body of Knowledge Fulfillment*

Table 2 summarizes the following three-step model for validating path 1:

*Validation of path 1 to BOK fulfillment relies on ABET, an approved outside entity, and licensing boards.*

**Step 1:** ABET, Inc., validates the fulfillment of the B component of the BOK (see Figure 3) through the formal accreditation processes of the Engineering Accreditation Commission of ABET (EAC/ABET). Specifically, the criteria for an accredited bachelor's degree in civil engineering contains the appropriate language that validates the portion of the BOK fulfilled through the bachelor's degree. This language is included within the General Criteria for Baccalaureate Level Programs and the Program Criteria for Civil and Similarly Named Engineering Programs of the *Criteria for Accrediting Engineering Programs.*

**Step 2:** Accreditation of the civil engineering master's degree by the EAC/ABET is not relied upon in path 1, even though it is an acceptable and efficient means of validating the M/30 component of the BOK. More generally, an approved outside entity (AOE) validates the fulfillment of the M/30 component of the BOK (see Figure 3) outside of the normal EAC/ABET accreditation process. Because the B components included in the baccalaureate-level general criteria and the program criteria have been validated by the EAC/ABET, the validation by an AOE is limited to the M/30 component. Alternative approaches for validating the M/30 component of the BOK were explored by the CAP$^3$ Fulfillment and Validation Committee.[15] The National Council of Examiners for Engineering and Surveying (NCEES) is also actively working to define AOEs in conjunction with revisions to the NCEES Model Law and the NCEES Model Rules. CAP$^3$ is confident that these efforts will confirm that validation of the M/30 component of the BOK by AOEs is viable—and will be more clearly defined in the near future.

**Step 3:** Historically, individual state licensing boards have validated an individual's completion of required prelicensure experiential development. In the short term, CAP$^3$'s validation model will continue to use individual state licensing boards for validating the E component of the BOK. Because the current experience guidelines used by state licensing boards are not based upon the BOK outcomes, the validation of the portion of the BOK fulfilled through prelicensure experience is incomplete. In February 2007, CAP$^3$ established an Experience Committee to conduct initial exploratory research on alternatives to handling experience guidelines for entry into the civil engineering profession. This committee completed its report[16] in July 2007. (See the "Experience Guidelines" section later in this chapter for additional discussion of the experience component.)

Table 2. Validation of path 1 to body of knowledge fulfillment.

| $B^{ABET}$ + $(M/30)^{Validated}$ & E | | | |
|---|---|---|---|
| *Step* | *Validation* | *Component validated (from Figure 3)* | *Validation entity* |
| #1 | $B^{ABET}$ | B | ABET, Inc. |
| #2 | $M^{Validated}$ or $30^{Validated}$ | M/30 | Approved outside entity |
| #3 | E | E | State licensing board |

## ***Validation of Path 2 to Body of Knowledge Fulfillment***

Table 3 summarizes the following two-step model for validating path 2:

*Validation of path 2 to BOK fulfillment is based on ABET and licensing boards.*

**Step 1:** Accreditation of the bachelor's degree by ABET is not relied upon in path 2, even though perfectly acceptable. Instead, ABET validates the fulfillment of both the B and the M/30 components of the BOK (see Figure 3) through an EAC/ABET-accredited master's degree. Specifically, the criteria for an accredited master's degree in civil engineering contains all of the appropriate language for validating the portion of the BOK fulfilled through the bachelor's degree (which may not be accredited by EAC/ABET) and the master's degree (which must be accredited by EAC/ABET). This language is included within the General Criteria for Masters Level Programs in the Criteria for Accrediting Engineering Programs. The master's level general criteria state:

> *The criteria for masters level programs are fulfillment of the baccalaureate level general criteria, fulfillment of program criteria appropriate to the masters level specialization area, and one academic year of study beyond the basic level. The program must demonstrate that graduates have an ability to apply masters level knowledge in a specialized area of engineering related to the program area.*

This wording was crafted to validate both the B and M/30 components of the BOK—for all individuals earning an EAC/ABET master's degree in civil engineering.

Table 3. Validation of path 2 to body of knowledge fulfillment.

| B + $M^{ABET}$ & E | | | |
|---|---|---|---|
| *Step* | *Validation* | *Component validated (from Figure 3)* | *Validation entity* |
| #1 | $M^{ABET}$ | B<br>M | ABET, Inc. |
| #2 | E | E | State licensing board |

**Step 2:** This step is identical to the third step of the previous validation model. It uses the individual state licensing boards for validating the E component of the BOK.

In summary, two models are available for validating an individual's attainment of the BOK. Path 2 offers the opportunity for a strong student with a non-accredited engineering degree, or with a science or mathematics degree, to fulfill the civil engineering BOK. It is simpler—relying solely on existing accreditation and licensing organizations. Path 1, while more complex, ensures the most flexible path for BOK fulfillment.

## The Vision for Civil Engineering in 2025 and the Body of Knowledge: The Foundation of the Policy Statement 465 Master Plan

### *Introduction*

*The master plan for implementation of ASCE Policy Statement 465 is founded on the BOK.*

Earlier sections of this chapter addressed the question, how is the BOK fulfilled by tomorrow's aspiring civil engineer? This section addresses the different and more strategic question of how the BOK is used by ASCE to change the prerequisites for entry into the professional practice of civil engineering. The concise answer is that the BOK is used as the starting point of the entire ASCE master plan[17] to "raise the bar" for entry into professional practice. In other words, the BOK is the foundation of the ASCE master plan to implement PS 465.

### *Overview of the Master Plan*

The master plan must be understood in order to appreciate how the BOK is used by ASCE to implement PS 465. ASCE's complex, multidimensional and integrated master plan, as

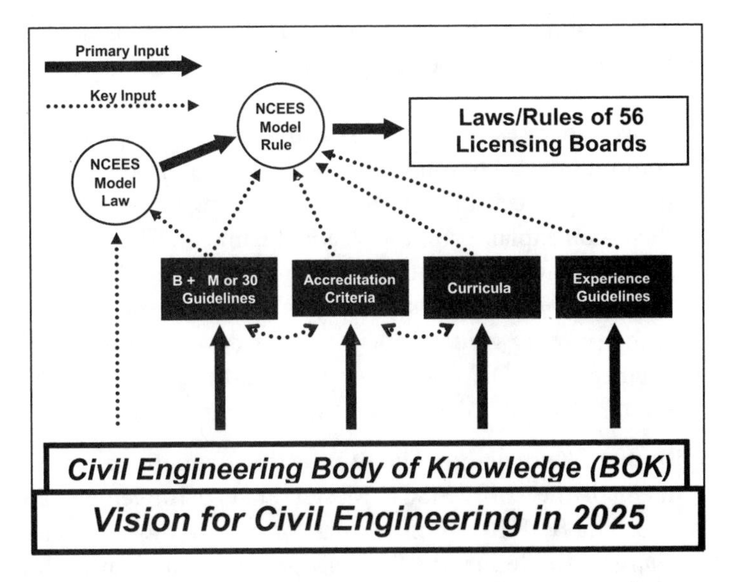


Figure 4. ASCE's master plan for implementing Policy Statement 465 builds on the Vision for Civil Engineering in 2025 and the body of knowledge.

shown in Figure 4, is based entirely on the BOK. In executing this plan, ASCE will use the BOK to change the educational and licensure processes of the civil engineering profession, including:

- accreditation criteria of engineering programs,
- university curricula,
- on-the-job education and training of engineer interns,
- NCEES Model Law and Model Rules, and,
- ultimately state, District of Columbia, and U.S. territory laws and regulations governing the licensure of practicing professional engineers.

The various elements of this plan are discussed in the following section.

### *The Master Plan: Work Products and Committees*

The work products associated with this master plan, as well as the committees working on these products, are briefly explained below. These products are also being integrated into the ASCE strategic planning process. (For additional details about the work products of the $CAP^3$ committees, visit www.asce.org/raisethebar.)

**The Vision for Civil Engineering in 2025:** As explained in Chapter 1 of this report, ASCE intends that the document *The Vision for Civil Engineering in 2025*[1] will guide policies, plans, processes, and progress within the civil engineering community and beyond, worldwide. Reform in the education and prelicensure experience of civil engineers supports the vision explained in the Vision document. The vision is strategic, future-oriented, comprehensive, and aspirational—characteristics shared by the BOK. As such, Vision is the primary guiding document for developing the civil engineering BOK and the logical starting point for the master plan guiding the realization of ASCE PS 465.

*The refined BOK described in this report strengthens the master plan's foundation.*

**Body of Knowledge:** The BOK is the foundational document of the entire ASCE Master Plan to Implement PS 465. The BOK1 Committee was formed in May 2002 and charged to define the knowledge, skills, and attitudes needed to enter the practice of civil engineering at the professional level. The committee published the *Civil Engineering Body of Knowledge for the 21st Century: Preparing the Civil Engineer for the Future* in January 2004.[3] A new BOK committee (the BOK2 Committee) was appointed in October 2005 to prepare and publish this refined second edition of the BOK. (See Appendix E for the details regarding the BOK2 Committee and many others who contributed to the preparation of this report.)

**Curricula:** As BOK1 was nearing completion in late 2003, CAP[3] organized a group consisting primarily, but not exclusively, of civil engineering faculty to, first, determine the current status of civil engineering education with respect to the formal educational component of the first edition of the BOK and, second, determine the nature of change necessary to support the formal educational expectations of the BOK. The Curricula Committee of CAP[3] was given this charge in September 2003.

The Curricula Committee, which completed its report[18] in December 2006, concluded that:

- The original BOK "is not accomplished within current (undergraduate) civil engineering curricula."
- The formal education component of the original BOK, "except for the outcome regarding technical specialization, can be included in the undergraduate curriculum."
- "Specialized technical knowledge is best accomplished in a postgraduate program of study."

The National Academy of Engineering (NAE) reached similar conclusions in its 2005 report.[5] NAE report recommendations include the following: "The B.S. degree should be considered as a pre-engineering or 'engineer in training' degree. Engineering programs should be accredited at both the B.S. and M.S. levels so that the M.S. degree can be recognized as the engineering 'professional' degree."

*ASCE and the National Academy of Engineering are aligned.*

A new curricula-related committee, the BOK Educational Fulfillment Committee, was formed in late 2007 and charged with reviewing the second edition of the BOK and associated work products that affect the formal educational process.

**Accreditation Criteria:** Using the BOK1 as its primary reference, the CAP³ Accreditation Committee crafted new civil engineering program criteria and new master's-level criteria.[19] These criteria, which were approved by the ABET Board of Direction in November 2007, will be used for the first time during the 2008–2009 accreditation cycle.

*New BOK-consistent accreditation criteria will be applied for the first time during the 2008-2009 accreditation cycle.*

The Accreditation Committee also drafted the ASCE commentary to the criteria.[20] This draft commentary provides civil engineering program evaluators with guidelines for applying the new criteria and provides civil engineering faculty with recommended measures to ensure full, robust implementation of the BOK1. Using these new criteria and this commentary, civil engineering programs will be accredited within the context of the *Civil Engineering Body of Knowledge for the 21st Century: Preparing the Civil Engineer for the Future.*

**Experience Guidelines:** As discussed earlier in this chapter, pre-licensure experience is essential to fulfillment of the BOK. Using the BOK2 as its basic reference, a new BOK Experiential Fulfillment Committee will review, prepare, and publish guidelines to assist the engineer intern in achieving those outcomes identified for partial fulfillment through on-the-job education and training. The work products of this committee will be posted as soon as they are available at www.asce.org/raisethebar.

*Experience guidelines are being developed.*

**B + M or 30 Guidelines:** Based upon the fulfillment models presented in the first edition of the *Civil Engineering Body of Knowledge for the 21st Century,* the Fulfillment and Validation Committee of CAP³ began work in September 2004. This committee explored how alternative education providers, other than universities, could provide creditable postgraduate engineering education. It investigated alternative education providers that could provide academic rigor and individual assessment comparable to traditional universities. This committee also addressed how to validate the "+30" portion of the BOK. The committee's 2005 report[15] influenced the contents of the Model Law and Model Rules proposed by the NCEES.

Several committees worked, or will be working, on elements related to the B + M/30 guidelines to supplement the report of the Fulfillment and Validation Committee. For example:

- The September 2005 report[9] of the Levels of Achievement Subcommittee of the Curricula Committee of CAP[3] recommended the portion of the prelicensure level of achievement to be fulfilled through the bachelor's degree, the master's degree or equivalent, and prelicensure experience. The BOK2 Committee applied this type of allocation to its recommended 24 outcomes.
- The Licensure Committee of CAP[3] is working with the NCEES Bachelor's +30 Task Force to develop definitions for approved credits and approved course providers for inclusion in the NCEES Model Rules.
- The CAP[3] BOK Educational Fulfillment Committee (BOKEdFC) will be reviewing the work products of the BOK2 Committee—particularly the BOK2 rubric and "where" each outcome should be fulfilled.
- The CAP[3] BOK Experiential Fulfillment Committee (BOKExFC) will be preparing guidelines to assist engineer interns in achieving outcome fulfillment through on-the-job education and training.

*The 2006 changes in the NCEES Model Law support BOK fulfillment.*

**Model Law, Model Rules, and State Licensing Laws/Rules:** ASCE PS 465 states that the attainment of a BOK would be accomplished through the adoption of appropriate engineering education and experience requirements as a prerequisite for licensure. In other words, ASCE has unequivocally decided that implementation of ASCE PS 465 requires that the BOK be tied to the profession's licensure process.

The National Council of Examiners for Engineering and Surveying (NCEES) has been studying this issue for the last six years. As a result, the *Engineering Licensure Qual-*

*ifications Task Force (ELQTF) Report* was released in 2003. The report was prepared in collaboration with 11 other engineering societies, including the National Society of Professional Engineers (NSPE). The report recommended that additional education be considered as part of the requirements for licensure in the future. The task force report was issued to the Licensure Qualifications Oversight Group (LQOG) of the NCEES. The LQOG performed additional research and considered the conclusions and recommendations of the ELQTF from an NCEES and regulatory perspective. In 2005, the concept of the bachelor's plus was approved by the NCEES and subsequently, in 2006, the Uniform Policies & Legislative Guidelines (UP&LG) committee drafted language to modify the Model Law to include the bachelor's + 30 concept beginning in 2015.

In September 2006, delegates at the NCEES annual meeting approved the modifications to the Model Law requirements to require additional education for engineering licensure. This future-focused change to the NCEES Model Law was confirmed by delegates at the August 2007 NCEES annual meeting.

The approved language states that an engineer intern with a bachelor's degree must have an additional 30 semester credits of acceptable upper-level undergraduate or graduate-level coursework from approved providers in order to be admitted to the Principles and Practice of Engineering (P.E.) Examination. A master's degree or Ph.D. from an approved institution would also qualify. The change, to be effective in 2015, is a recommendation to each of the 55 licensing jurisdictions,[21] which individually will have to modify their state laws and/or rules to reflect the new NCEES Model Law and Model Rules. Up-to-date information concerning changes to the NCEES Model Law and Model Rules can be found at www.ncees.org/licensure/licensure_exchange/.

In summary, this section describes how the BOK is being used by ASCE as the foundation for the entire master plan to implement PS 465. The BOK affects the work products of every constituent committee of CAP³.

## Other Ways the Body of Knowledge Could Be Used

*The BOK is relevant to many and varied civil engineering stakeholders.*

When well crafted, a profession's BOK speaks to all segments of the profession. While the messages may differ among the various segments, all can view the BOK as common ground. The BOK is a foundation on which a profession's members study and build careers, carry out their responsibilities, and pursue opportunities. Consider the relevance of BOK2 to various members of and stakeholders in the civil engineering community. The BOK2:

- offers *prospective civil engineering students,* and their *parents, teachers, counselors,* and *advisors,* a glimpse of the importance of civil engineering and the breadth of opportunities offered;
- assists *civil engineering and other faculty* in designing curricula, creating and improving courses, and teaching and counseling students;
- offers *researchers* ideas on future directions of civil engineering and related technical needs and defines the knowledge, skills, and attitudes that should be offered by students seeking to engage in research;
- provides *civil engineering students* and *engineer interns* with a framework against which they can understand the purpose, measure the progress, and plan the completion of their studies and prelicensure experience;
- gives *ABET leaders* a basis for developing appropriate accreditation criteria;
- informs *employers* what they can expect in terms of basic knowledge, skills, and attitudes possessed by civil engineering graduates;
- suggests to *practitioners* their role, in partnership with individual engineer interns prior to licensure, in helping them attain the additional levels of achievement needed to enter the practice of civil engineering at the professional level;
- provides *licensing boards* with confidence that the formal education and prelicensure experience of civil engineers will meet the engineering profession's responsibility to protect public safety, health, and welfare; and
- encourages *specialty certification boards* to build on the prelicensure BOK in defining their desired mastery level of achievement.

# CHAPTER 4

# Guidance for Faculty, Students, Engineer Interns, and Practitioners

*In a time of drastic change, it is the learners who inherit the future. The learned usually find themselves equipped to live in a world that no longer exists.*
Eric Hoffer, self-taught social philosopher

## Introduction

*BOK guidance is offered for faculty, students, engineer interns, and practitioners.*

ASCE's 1995 Civil Engineering Education Conference[22] recommended action in four areas: professional degrees, integrated curriculum, faculty development, and practitioner involvement. ASCE Policy Statement 465, in a broad sense, addresses all four. For example, the "what" and "how" of this BOK report—which are the themes of Chapter 2 and Chapter 3—relate directly to "professional degrees" and "an integrated curriculum." The "who" forms the theme of this chapter, because the chapter addresses "faculty development" and the complementary topic of "practitioner involvement." Also discussed within the chapter are students and engineer interns, both of whom are the principal BOK learners and beneficiaries.

PS 465 and its foundation, the BOK, are intended to reform the education and prelicensure experience of tomorrow's civil engineers. The resulting changes will naturally raise concern for some individuals. Accordingly, the chapter presumes that gradual implementation of the BOK will cause many affected members of the civil engineering community to be receptive to guidance. As noted, these members are very likely to include faculty members, students, engineer interns, and practitioners—especially those who coach and mentor engineer interns.

Any guidance that might be offered to those individuals must be credible. Accordingly, a variety of accomplished professionals drawn from academia and practice and from the private and public sectors were invited to offer guidance ideas for one of the four "who" groups. The "Contributors to Special Tasks" section of Appendix E lists the individuals who kindly accepted the BOK2 Committee's invitation, drew on their knowledge and experience, and provided guidance. Their ideas, edited for brevity and consistency, are the substance of this chapter. This chapter is written partly in second person to stress the guidance objective—that is, the chapter speaks to you as an interested faculty member, student, engineer intern, or practitioner. The committee is confident you will find, within this chapter, insights and advice applicable to your particular situation.

## Guidance for Faculty

As noted by ABET, "The faculty is the heart of any education program." You, as a civil engineering faculty member, are among the first representatives of the profession who most future civil engineers encounter and, as such, serve as their first professional role models. Therefore, in a very real sense, the future of civil engineering is dependent upon you and your colleagues. Earning and maintaining the respect of the public as professionals and leaders require that future engineers be well prepared.

***Respecting the crucial role of teachers, the committee contemplated the ideal civil engineering faculty of the future and their even more important role.***

The BOK2 Committee contemplated the ideal civil engineering faculty of the future and their important role in educating future generations of civil engineers. Who should the faculty be, as individuals and collectively? What will enable them to be successful in facilitating the accomplishment of the BOK? How can they present themselves with a professional attitude and as positive role models? How can they develop graduates who can creatively apply concepts to recognize, define, and solve engineering problems? What are the characteristics required of educators to aid them in motivating and guiding students toward the achievement of the BOK? And how can they do all this while also supporting their own professional development as well as the specific mission of their academic institution?

Interactions between students and teachers can affect education positively or negatively. High-quality teaching and mentoring can have a dramatic effect on retention. The two primary reasons cited by students as to why they switch from engineering to another major are poor teaching and inadequate advising. A third reason cited by the students is curriculum overload.[23]

*There is a need to strengthen civil engineering teaching.*

On the whole, frequent student interaction with faculty has positive effects on student development, involvement, and retention. However, one study found that greater interaction with faculty may not have the same positive effect.[24] In other words, poor attitudes and lack of professionalism by faculty have a dramatically negative effect on students. The need for high-quality and effective faculty to teach the civil engineering BOK is clear.

The BOK2 Committee reconfirmed the following four characteristics of the model full- or part-time civil engineering faculty member as presented by the BOK1 Committee:

*Model full- or part-time faculty should be scholars and effective teachers, have relevant practical experience, and serve as positive role models.*

- ***Scholars:*** Those who teach the civil engineering BOK should be scholars. You should acquire and maintain a high level of expertise in subjects that you teach. Being a scholar mandates that engineering faculty be lifelong learners, modeling continued growth in knowledge and understanding. Additionally, being a scholar requires that you be engaged in scholarly activities supporting your educational activities and your professional area(s) of practice.
- ***Effective Teachers:*** Student learning is optimal when you and other faculty members effectively engage students in the learning process. The development of engineering faculty as effective teachers is critical for the future of the profession. Faculty should be expected to gain pedagogical training through internal programs within their home institution or external programs offered through various professional organizations. An excellent example is ASCE's Excellence in Civil Engineering Education (ExCEEd) program.
- ***Have Relevant Practical Experience:*** Engineering is a profession of practice, so the education process must integrate this experiential component to be successful. You should have an appropriate level of relevant practical experience in the engineering subjects that you teach. Faculty have difficulty being passionate about the subjects they teach or fully communicating the relevance of the topic to students without appropriate experience.

- ***Positive Role Models:*** Regardless of personal desires or choice, every civil engineer who is in contact with students serves as a role model for the profession. You should be aware that students are viewing you in that light. The ideal civil engineering faculty member should present a positive role model for the profession. Students should be able to both relate to and follow these role models and be put on a path toward becoming successful engineers in their own right.

### Be a Scholar

*This use of the term "scholar" goes far beyond the traditional, restrictive view of scholarship as basic research.*

This use of the term "scholar" goes far beyond the traditional, restrictive view of scholarship as basic or applied research. Instead, the committee adopted the more inclusive view of scholarship espoused by Ernest L. Boyer in his seminal work, *Scholarship Reconsidered: Priorities of the Professoriate.*[25]

Boyer recognized that knowledge can be acquired through research, synthesis, practice, and teaching. He defined the four corresponding functions of scholars as the scholarship of teaching, the scholarship of discovery, the scholarship of integration, and the scholarship of application. Scholars are true lifelong learners, continually acquiring knowledge. The four forms of scholarship are explained as follows:

- The ***Scholarship of Teaching*** comprises developing examples, analogies, and images that form the bridge between the teacher's understanding and the student's learning. It clearly satisfies the expectation of expertise in faculty, as it requires that faculty also be learners, always extending their own knowledge and understanding.
- The ***Scholarship of Discovery*** is the familiar, disciplined, investigative research. It enhances the meaning of the academy itself, discovering basic knowledge and continuing the intellectual climate of the university.
- The ***Scholarship of Integration*** makes connections within and between disciplines. As the master integrators of the infrastructure, civil engineers are rightly very interested in the synthesis of such multidisciplinary work.
- The ***Scholarship of Application*** is the professional activity of applying new knowledge to consequential problems. The civil engineer has clear ties to this scholarship, seeking to solve the challenges and problems of our infrastructure.

To further define scholarly work, consider its standards. Glassick[26] defines those standards as clear goals, adequate preparation, appropriate methods, significant results, and reflective critique.

By pursuing a mix of the four types of scholarship to achieve personal and institutional missions and goals, faculty and institutions—clear in their distinctive mission—will provide a more diverse graduate to the profession and will add to the richness of the education of the civil engineering profession. As stated by Boyer,[25] who advocated diversity with dignity:

> *Broadening scholarship has implications not only for individuals but for institutions, too. Today's higher education leaders speak with pride about the distinctive missions of their campuses. But such talk often masks a pattern of conformity. Too many campuses are inclined to seek status by imitating what they perceive to be more prestigious institutions. We are persuaded that if scholarship is to be enriched, every college and university must clarify its own goals and seek to relate its own unique purposes more directly to the reward system for professors.*

### *Teach Effectively*

Numerous studies indicate that student learning is enhanced when engineering faculty are effective and enthusiastic teachers. Under the current system of studies, civil engineers do not typically become effective teachers simply by advanced study leading to a Ph.D. Furthermore, civil engineers do not typically become effective teachers via experience obtained through practicing civil engineering. Appropriate teaching pedagogy and education training are critical to enhancing the effectiveness of faculty in creating excitement for learning by students.

*Appropriate teaching education and training are critical to enhancing the effectiveness of faculty in creating excitement for learning by students.*

Effective teaching is a challenging task, requiring expertise in the topic to be taught; effective two-way communication with students; an ability to promote clear, complex, and complete understanding; an awareness of learning styles; and an ability to relate to students in ways both positive and inspirational. You, as the teacher, must motivate students by active involvement in the individual student's personal learning process. Student learning is enhanced when the teacher is highly effective.

*The teacher who demonstrates interpersonal rapport with students will show interest in students as individuals, interest in students' learning, and openness to students' preferences about classroom procedures, policies, and assignments.*

*The effective teacher's skill comes from creating intellectual excitement in and interpersonal rapport with the students in a variety of classroom settings.*

In his book *Mastering the Techniques of Teaching,*[27] Joseph Lowman provides a two-dimensional model of effective college teaching. This model is shown in matrix form in Table 4. Lowman notes that the effective teacher's skill comes from creating intellectual excitement in and interpersonal rapport with the students in a variety of classroom settings. Although outstanding abilities in either dimension can result in adequate teaching for some students and success in certain kinds of classes, both dimensions are required for excellence in teaching. Faculty willing to learn and develop can improve in either or both dimensions.

A teacher's ability to create intellectual excitement (as shown by the rows of Table 4) has two components: clarity of presentation and stimulating emotional impact in the student. Lowman notes clarity deals with what the teacher presents. Stimulating emotional impact stems from the way it is presented.

Intellectual excitement is apparent from a teacher's technical expertise, organization, clarity of communication, engaging presentation, and enthusiasm. Descriptors associated with teachers who are skilled at developing intellectual excitement include knowledgeable, organized, interesting, humorous, clear, inspiring, and enthusiastic.

*The teacher who demonstrates interpersonal rapport with students will show interest in students as individuals, interest in students' learning, and openness to students' preferences about classroom procedures, policies, and assignments.*

The development of *interpersonal rapport* (as shown by the columns of Table 4) stems from the teacher's ability to effectively communicate with students in ways that increase their motivation, enjoyment, and independent learning. The teacher who demonstrates interpersonal rapport with students will show interest in students as individuals, interest in students' learning, and openness to students' preferences about classroom procedures, policies, and assignments. Terms used to describe teachers who demonstrate interpersonal rapport include concerned, encouraging, caring, helpful, challenging, available, and approachable.

Table 4. Lowman's two-dimensional model of effective college teaching is based on interpersonal rapport and intellectual excitement.

| *Intellectual Excitement* | *Interpersonal Rapport* | | |
|---|---|---|---|
| | *Low* | *Moderate* | *High* |
| High | 6. Intellectual authority | 8. Exemplary lecturer | 9. Complete exemplary |
| Moderate | 3. Adequate | 5. Competent | 7. Exemplary facilitator |
| Low | 1. Inadequate | 2. Marginal | 4. Socratic |

Lowman[27] suggests that you, as a teacher who desires to improve, should focus on intellectual excitement first, then interpersonal rapport. He says, "Unless traditional teaching skills are mastered first, structural inventions are unlikely to lead to exemplary instruction or optimal student learning." As evident in Table 4, development of intellectual excitement achieves higher levels of effectiveness for single step change.

Skills required to be an effective teacher can be learned. The committee recommends that all prospective engineering faculty members actively prepare to be effective teachers by means of pedagogical education and training courses as early in their career as possible and preferably before they teach civil engineering students. Many university campuses offer teaching programs. Some are within the engineering college while others are housed elsewhere, such as in the education school. ASCE's ExCEEd Teaching Workshop (ETW) provides a proven model for how to teach faculty to teach.[28-34] In this workshop, a detailed structure for success in the classroom is provided for civil engineering faculty.

*Prospective engineering faculty members should actively prepare, by means of education and training courses, to be effective teachers—preferably before they teach civil engineering students.*

### *Gain Relevant Practical Experience*

Boyer[25] states, "...teaching begins with what the teacher knows. Those who teach must, above all, be well informed and steeped in the knowledge of their fields." To have a solid mastery of their field, the BOK2 Committee maintains that civil engineering faculty should have relevant practical experience in the subjects they teach. A good guide of minimum relevant practical experience required is you must be able to both perform the engineering being taught as well as critique and judge the relative merits of alternative solutions in the context of project-specific constraints. Another guide is that each individual faculty member need not have experience in all or even multiple areas of civil engineering practice, but the department's faculty considered as a team should collectively possess sufficient practical experience relevant to those areas taught in that department.

Relevant practical experience may be gained as an employed engineer for a consulting firm, industry, or government agency.

*Relevant practical experience can be gained in many ways, such as consulting and research.*

Alternatively, relevant practical experience may be gained, or supplemented, through consulting on engineering projects while serving as faculty members. Relevant practical experience can sometimes be gained through a faculty member's research and outreach activities, depending on the specifics of these efforts. In performing research, working with practice-oriented agencies—for example, state and federal departments of transportation, industry associations, or individual companies—to develop new methods or technologies and working to advance the new methods or technologies into practice may constitute relevant practical experience.

*Faculty members with relevant practical experience will be better prepared to mentor students as they prepare to enter the engineering workforce.*

The benefits of relevant practical experience should include knowledge of the day-to-day operations of engineering projects, including many of the business aspects not always included in traditional civil engineering curricula but now being recommended as part of the BOK. Also, you and other faculty members with relevant practical experience will be better prepared to mentor students as they prepare to continue their education at the graduate level and/or enter the engineering workforce.

*While the majority of faculty will be full-time engineering educators, some should be part-time, leading-edge practitioners.*

Students who aspire to practice civil engineering at the professional level will benefit from a heterogeneous group of faculty, ranging from some who are fully engaged in academia to others who are fully engaged in the traditional practice of civil engineering. While the majority of faculty will be full-time engineering educators, some should be part-time, leading-edge practitioners.

Potential practitioner participants should meet the same criteria as the full-time faculty as described in this section—namely, scholarship, teaching effectiveness, and positive role modeling. Practitioner faculty might teach entire courses or co-teach with full-time faculty.

### *Serve as a Positive Role Model*

For many students, the first civil engineer they meet is a civil engineering faculty member. Beyond that, every civil engineering teacher continues to serve as a role model for the profession throughout the student's academic career. Those learning the BOK will look to you and other civil engineering faculty for appropriate knowledge, skills, and attitudes desired of civil engineers.

*Students will hold each civil engineering faculty member as an example to emulate or as an example to reject.*

Whether or not a faculty member desires to be a role model for the engineering profession is irrelevant. In every case students will hold each civil engineering faculty member as an example to emulate or as an example to reject. The ideal civil engineering faculty member will be a positive role model for the profession.

Civil engineering faculty should personally strive for and be encouraged to gain relevant experience to supplement their academic knowledge and increase their effectiveness in the classroom. Furthermore, when appropriate, civil engineering faculty should obtain professional licensure. For those faculty members who teach civil engineering design courses, relevant design experience in the topics they teach is necessary and professional licensure holds particular relevance. Others are encouraged to obtain licensure to emphasize its importance to students.

*Faculty are encouraged to obtain professional licensure, partly to emphasize its importance to students.*

Faculty who are lifelong learners will infuse the continued thirst for new solutions to the challenges within the profession. Effective teachers will stimulate the curiosity of their students and will exemplify the knowledge, skills, attitudes, and behaviors that best reflect the civil engineering profession.

### Balance Teaching with Other Responsibilities

Tomorrow, as is true today, faculty will be expected to do much more than be effective teachers as described in this section. Depending on the specific mission of their academic institution, faculty members will be expected to provide a quality educational experience for their students and contribute to other departmental or institutional goals. In most universities, research and service complete a tripartite mission. Faculty must work with their department chair to understand the specific expectations that are unique to their program.

*Faculty will need to balance teaching, research, and service in accordance with the expectations of their specific institutions.*

While there are no set paths or criteria for successful civil engineering faculty, being part of a program with a mission that is compatible with one's own professional interests and goals is the key. Faculty members—especially new ones—should seek programs such that their own professional interests and aspirations are supportive of the program's mission. Beyond this, you and your faculty colleagues should share a common sense of commitment to the civil engineering students you teach, including a commitment to the attributes presented in this section.

The service component of a faculty member's responsibility will include service to the civil engineering profession, both on campus (for example, mentoring and supporting student chapter activities of engineering organizations) and also off campus (for example, by volunteering services in professional societies), where the skills and experience of a faculty member help shape the preparation of the civil engineer of the future.

### *Summary*

Model civil engineering faculty are scholars and effective teachers with an appropriate level of relevant practical experience and are positive role models for the profession. These are desirable traits for those who will motivate and guide 21$^{st}$-century civil engineers.

## Guidance for Students

### *Understand the Vision*

*The 2025 civil engineering vision coupled with the BOK will guide students.*

Review the vision for civil engineering in 2025 as described in Chapter 1 of this report. The civil engineering profession knows where it is going and invites you to join the journey. You can help achieve the vision by fulfilling the BOK and entering the practice of civil engineering at the professional level. The vision, coupled with the BOK outcomes and levels of achievement, should provide you with a framework within which you can understand the purpose and measure the progress of your education, prepare to move into your internship, and, ultimately enter the practice of civil engineering at the professional level.

### *Utilize Campus Resources*

As a civil engineering student, you will be faced with challenges in and outside of the classroom. For example, you may fail an examination, receive a low grade in a course, have difficulty understanding certain fundamentals, or encounter problems financing your education. Fortunately, you are likely to be surrounded by many and varied resources typically available on campuses. Personal examples are friends, professors, advisors, and counselors. Your campus is likely to have programs, centers, and offices that can assist you with time management, writing, studying, tutoring, computing, financial aid, part-time work, and summer and permanent employment. Draw on selected resources, depending on your needs, so that you continue to move forward in your formal education.

### *Actively Participate in Campus Organizations*

You can move toward fulfillment of outcome 16 (communication), outcome 20 (leadership), and outcome 21 (teamwork) by active, as apposed to passive, participation in one or more campus organizations. You could choose from the student chapters of such engineering organizations as ASCE, NSPE, the Society of Women Engineers, the Society of Hispanic Professional Engineers, and the National Society of Black Engineers. However, you can also learn about communication, leadership, and teamwork by being actively involved in such campus-wide activities and groups as student government, service clubs, sports teams, a student newspaper, and sororities and fraternities. Consider your active participation in such groups as these as an opportunity to serve while enhancing your knowledge, skills, and attitudes.

*Be actively, as opposed to passively, involved in at least one campus organization.*

### *Explore International Programs*

The explanation for outcome 19 (globalization), offers you this advice: "Engineers will need to deal with ever-increasing globalization; and find ways to prosper within an integrated international environment; and meet challenges that cross cultural, language, legal, and political boundaries...." Given the impact of globalization on engineering, you should at least explore participating in an international study program.

*Students should, at minimum, explore participating in an international study program.*

Many are available and they both literally and figuratively cover the globe. These programs typically involve a semester or so of study at a university in another country along with such other learning opportunities as summer travel and/or work. While participation in an international program may extend the length of your formal education, that is likely to be a small cost relative to the added benefits.

### *Seek Relevant Work Experiences*

You can apply and augment your classroom and laboratory learning during your formal education by finding relevant work experience. Applying what you have learned deepens your understanding of the material and demonstrates the relevance of your ongoing formal education. Compensation for this work can also help to finance your education. Work options include part-time employment with a local engineering organization, summer employment, internships, and cooperative education.

*You are now beginning to create a reputation, good or bad, that will follow you throughout your career.*

### *Protect Your Reputation*

Craftsmen are judged primarily by the objects they create. Engineers, in contrast, are judged primarily by the credibility of their advice. Most of the clients and others you will eventually serve will not be able to fully judge the technical and other advice you offer. However, they will be aware of and be able to judge your reputation and use that to value and trust—or devalue and mistrust—you.

You may think that this scenario is years away for you—that it is not relevant now while you are in school. However, your reputation as a professional is beginning now, while you are a student. Years from now, individuals who are now students, faculty, and staff will recall what you said and did. Cherish, protect, and enhance your reputation by what you say and do. Tell the truth. Keep your word. Be careful what you write in e-mails, memoranda, letters, and reports. Give credit when using ideas, data, and information developed by others. Stated differently, recognize the necessity of fulfilling outcome 24 (professional and ethical responsibility).

### *Prepare Yourself for an Ever-Changing World*

Ancient Romans achieved an astonishing level of civil engineering excellence. Their works included extensive and complex viaduct and bridge structures. An example is the Pont du Gard in southern France, a towering structure composed of three tiers of arches that still stands 2,000 years after it was designed and constructed.

The civil engineering profession has come a long way since then. You are learning about an array of sophisticated tools and complex materials, including computer-aided drafting and design (CADD), digital models, sustainable design, analytical testing apparatuses, and composites. Just as today's practice is much different from yesterday's, so will tomorrow's practice—your practice—be much different from today's. The BOK, built on 4 foundational outcomes and having a broad and deep superstructure of 11 technical and 9 professional outcomes, will help you adjust to inevitable changes and prepare you to lead some of them. Furthermore, various books and other materials[35] are available to help you successfully complete your studies and proactively move into employment.

### *Find the Right First Job*

*Choose your employer carefully—especially your first employer.*

As your formal education draws to a close—whether it results in earning a bachelor's, master's, or other degree—you will naturally be thinking about employment. You are likely to consider many and varied factors in selecting an employer. Examples are compensation, benefits, location, computer resources, the functions you will perform, and the kinds of projects on which you will work. Choose wisely among the positions that will be available to you in the public and private sectors. In addition to, and perhaps more important than the preceding factors, are the following questions:

- Who will be your immediate supervisor? This is important because, early in your career, frequent interaction with him or her in a variety of settings will further influence your attitude toward the profession and the additional knowledge and skills you acquire. In a similar fashion, who will you work with? Choose your employer carefully.
- Does the organization have a positive culture—that is, does it value high expectations and provide support, partner with its personnel in their personal and professional development, insist on ethical behavior, and seek to be a leader among its peers?
- Is the potential employer aware of the BOK and your desire to complete its fulfillment so that you sit for the licensing examination? While you have the primary responsibility for fulfilling the BOK, you will benefit from a knowledgeable and supportive employer.

Best wishes as you enter this next critical and exciting phase of your career.

## Guidance for Engineer Interns

### *Self-Direct Your Life*

Prior to completion of formal education, your life has been largely directed by others—for example, parents, teachers, and coaches. They often told you what to do, how to do it, when to do it, and sometimes why to do it.

*Assume primary responsibility for your personal and professional development.*

Upon completion of education and entry into prelicensure practice, the situation changes—sometimes dramatically. The engineer intern moves from being directed primarily by others to being primarily self-directed. Assuming you fully embrace this transition to self-direction, including seeing it as part of the process of fulfilling the BOK, the "world is your oyster." The advice offered in this intern section assumes you are proactively becoming even more self-directed in your work life and beyond. More specifically, you are assuming primary responsibility for your personal and professional development while seeking support from your employer and others. The ultimate exercise in self-direction is to set goals and create plans by which to achieve them, which is the last topic in this intern section.

### *Continue Your Education*

*Your education has just begun.*

This advice ties directly to outcome 23 (lifelong learning). In addition, it builds on or advances essentially all of the other outcomes. If you temporarily ended your formal education with a baccalaureate degree, immediately prepare plans for earning a master's degree or approximately 30 semester credits of acceptable graduate-level or upper-level undergraduate courses. Earning the M/30 is an essential step in fulfilling the BOK. Regardless of your M/30 status, seek participation in continuing education. This can be an effective means of maintaining and advancing your knowledge and skills in very specific technical and nontechnical areas. Be willing and able to invest some of your time and money in formal and continuing education.

Civil engineers are fundamentally applied scientists. One indication of this is the inclusion in the BOK of foundational outcomes 1 and 2 (mathematics and natural sciences). Employers, clients, and stakeholders expect you to keep current so that they can benefit from the latest scientific discoveries and technological developments. In a recent report, the NAE[5] noted that the half-life of current engineering education is between two and four years.

*Seek breadth and depth of experiential and other learning.*

While you are likely to focus initially on scientific and technical topics, recognize that engineering encompasses nontechnical areas. Analysis and design are necessary, but engineering goes far beyond. Accordingly, seek both depth and breadth of experiential and other learning.

Read widely and eclectically, including articles, books, newspapers, and other publications that address a range of topics—technical, historical, economic, social, and contemporary. Consider the goal of reading a book a month. Subscribe to helpful e-newsletters. Another means of continuing your education lies right in front of you. Seize every opportunity for experiential learning in your day-to-day work as an intern. Seek a variety of assignments and increased responsibility.

Search for the most accomplished and respected individuals in your organization and strive to work for or with them. If your employer provides a formal mentoring program, consider participating first as a protégé and possibly later as a mentor. In the absence of a formal program, you may be able to informally find and benefit from a coach or mentor. Master the appropriate "knowns," especially those needed to contribute to your employer, while also trying to prepare yourself for the "unknowns."

### *Move Further Toward Licensure*

Recall that the BOK is defined in ASCE PS 465 as "the necessary depth and breadth of knowledge, skills, and attitudes required of an individual entering the practice of civil engineering at the professional level in the $21^{st}$ century." PS 465 goes on to explain that "entering the practice of civil engineering at the professional level" means licensure as a professional engineer (P.E.). Achieving licensure is one of the reasons to continue your education as advocated in the previous section.

Assuming full implementation of PS 465, you would need to demonstrate fulfillment of 24 outcomes to sit for the P.E. examination. As illustrated in Figure 3, while the foundation for all 24 outcomes was laid while you earned your bachelor's degree, and while three outcomes were or will be further fulfilled by earning your master's degree or approximately 30 credits, almost two thirds of the outcomes are to be further fulfilled during prelicensure experience. That's where you are right now.

*Proactively move toward licensure. Be wary of arguments against licensure.*

Some words of caution: Be wary of arguments—sometimes very self-serving—against licensure. Someone may say that you are working in an employment sector that is under the industrial exemption and therefore that you do not need a license. Will you always want to work in that sector? Others will oppose licensure because it results in having to pay higher compensation to licensed engineers. While they will not make that argument directly to you, if you are employed in their organization and are not licensed, you are likely to incur a penalty in compensation and opportunities.

Others will say that licensure is merely a shallow "prestige" credential and that your employment with them—and perhaps even others—is secure as long as you maintain your technical competence. After all, that's what really counts. But what if, someday, you want to start your own business—perhaps first as an individual proprietor and then later as the leader of a small and growing engineering firm? Can you exercise that option without a P.E.? Even if you never start your own firm, but choose instead to spend your professional career as an employee of an engineering organization, state laws require that the engineer in responsible charge of engineering work be licensed. Are you willing to relinquish this opportunity? That is very unlikely, so keep your options open by proactively seeking licensure.

### *Develop Horizontal Thinking*

Historically, engineering has tended to produce vertical thinkers. Picture engineering knowledge as a silo. Engineers are highly educated, trained, and skilled at knowing everything within their particular silo. Typically, the focus has been on the depth of a silo at the expense of knowledge outside the silo. Engineers have often had trouble embracing horizontal or lateral knowledge and thinking—that is, working effectively beyond the limits of their silo.

*Practice horizontal thinking so that you are connected to ideas and information not part of, but potentially related to, civil engineering.*

Many of the outcomes outlined in the BOK will require engineers to function horizontally—they will be stretched beyond the comfort of their silos. Fulfilling such outcomes as 3 (humanities), 4 (social sciences), 8 (problem recognition and solving), 10 (sustainability), 11 (contemporary issues and historical perspectives), 12 (risk and uncertainty), 16 (communication), 17 (public policy), 19 (globalization), 21 (teamwork), and 22 (attitudes) will enable you to further develop horizontal thinking.

Horizontal or lateral thinking connects you with ideas and information not a part of but potentially related to civil engineering. As a result, you will be better prepared—as part of intradisciplinary and multidisciplinary teams—to help identify and solve the complex problems of the future. The required innovation and creativity result, in part, from making personal and other connections along horizontal paths.

### *Volunteer in Community and Professional Organizations*

*Volunteer to serve the profession and your community.*

Civil engineering has a long and proud tradition of serving the public, as compensated professionals and as volunteers. By volunteering, engineer interns give of themselves and add value to their communities, as well as to professional societies. By giving of yourself you will further fulfill such outcomes as 16 (communication), 17 (public policy), 20 (leadership), 21 (teamwork), and 24 (professional and ethical responsibility).

Most communities have planning and zoning commissions, park and recreation boards, capital improvement committees, and similar entities. Positions in these groups are typically appointed by the city council or other elected officials. Groups like these will benefit from your technical and other contributions. You, as a young engineer, will have the opportunity to interact with a key segment of society—the nontechnical public—and become even more familiar with their concerns and more adept at communicating with the public. You are urged to proactively seek to serve as a volunteer—both you and the group you join will benefit.

Similarly, professional societies typically relevant to civil engineering—APWA, ASCE, ASEE, and NSPE, for example—rely heavily on volunteers. Consider offering to assist by speaking on behalf of a professional society at a local school, assisting a program committee, and offering to do whatever may be needed.

### *Reflect, Plan, and Act*

You, as an engineer intern, are likely to be very busy, fully engaged with work and your personal life. You will be doing useful work and finding satisfaction. As good as this may sound, there are dangers. You may be gaining experience, but perhaps too much of certain kinds. For example, rather than having four years of experience you could have one year of experience four times. Experience is wonderful, but too much of one kind of experience could diminish your rate of personal and professional development. Or you may be so focused on the tasks at hand that you fail to see the available range of professional opportunities and options. Accordingly, excessive focus on current tasks could lead to later regrets. Author and lecturer Og Mandino[36] observes that "... experience teaches thoroughly yet her course of instruction devours [our] years so that the value of her lessons diminishes with the time necessary to acquire her special wisdom."

*Take time to reflect on your personal and professional progress, set goals, create plans to achieve them, and then act.*

Therefore, take time as you move well into your internship to reflect on what you have experienced so far—your successes and your failures—and what, in the spirit of outcome 23 (lifelong learning) you have learned. Complement this retrospective exercise with a prospective effort—that is, develop a plan. This means setting goals for personal and professional development and identifying action items that will enable you to achieve those goals. Certainly include the remainder of your intern period in this goal-setting process. But also look beyond it, at least in a general way. For example, what do you want to accomplish by the age of 30, 40, and so on?

Then act on your action items. This should include selectively sharing goals with your supervisor, colleagues, friends, family, and others. You are likely to be pleasantly surprised to find that diverse individuals who care about you will assist you on achieving your goals.

## Guidance for Practitioners

### *Review the Body of Knowledge*

Faculty members, students, and engineer interns will naturally be familiar with the BOK because of their active participation in it. In contrast, at least initially, leaders and managers of private and public engineering organizations and practitioners in those organizations are less likely to have had direct contact with the BOK. Accordingly, you may want to review the BOK as described in this report. More specifically, consider studying Figure 3 and Appendix I and Appendix J, which collectively present the BOK rubric and an explanation of it. The rubric is the "heart" of the BOK—it is the "road map" used first by the student and now the intern to enter the practice of civil engineering at the professional level. Likewise, the rubric is a definitive statement of tomorrow's foundation of professional competence.

*Practitioners can assist the engineer intern in continuing the learning process while simultaneously benefiting the employer.*

As a practitioner, you can appreciate the importance of the young professional's early experience as an engineer intern in completing fulfillment of the BOK required for entry into the practice of civil engineering at the professional level—that is, licensure. You should also understand that, while attending to your various responsibilities, you can also help the intern continue his or her learning process in preparation for the licensing examination. BOK outcomes and the levels of achievement that must be fulfilled are broad and deep. You; the engineer interns you supervise, coach, or mentor; and your organization will benefit from that fulfillment process.

### *Provide a Professional Development Program*

*An employer-sponsored professional development program can assist engineer interns while strengthening the organization.*

The BOK2 Committee believes that a carefully crafted and periodically monitored professional development program can benefit both the organization—private or public—and the individual. Such a program can assist the engineer intern in completing fulfillment of the BOK. While BOK fulfillment is ultimately each engineer intern's responsibility, the availability of learning opportunities within the organization encourages and supports individual efforts. From an organizational perspective, a professional development program can enhance the collective knowledge, skills, and attitudes of the organization and, as a result, enhance its effectiveness.

An organizational professional development program could include various combinations of the following means of teaching and learning:

- Internal and external seminars, workshops, and conferences;
- Mentoring, tutoring, and coaching;
- Experiential learning resulting from planned participation in a variety of office and field functions;
- Active participation in professional and business societies; and
- Periodic reviews of individual goals and plans for and progress toward achieving those goals.

Of course the professional development program in your organization will be tailored to the immediate and long-term needs of your organization. It will be planned and implemented in recognition of profitability expectations in the private sector and budget constraints in the public sector. Hopefully, such a program will be viewed by you and others as an investment instead of a cost. And, as is the case with prudent investing, the professional development program will be monitored for organizational and individual effectiveness and continuously improved.

The effectiveness of your organization's professional development program is likely to be enhanced if it involves a partnership between individuals and the organization. Each party should invest. For example, if the organization offers a four-hour in-house workshop, it might be scheduled near the end of the work day. Two hours would be on "company time" and two hours on personal time. The suggested partnership approach offers two benefits. First, it reduces the impact of lost productive time. Second, it provides an opportunity for individuals to demonstrate their commitment to professional development by investing some of their time.

## Encourage and Support Experiential Learning

*Encourage the engineer intern to benefit from experiential learning by seeking a variety of assignments.*

The best way for the intern to learn about what the organization does and how it does it is to do it. Urge the engineer intern to seek a variety of assignments within the organization, within and beyond the context of formal projects. During their four-year internships, the interns could, with your encouragement and support, participate in a variety of project functions. Examples are assisting with proposals, field work, statistical analysis, formulating alternatives, estimating costs, seeking permits, writing reports, and making presentations.

Such experiential learning offers valuable lessons to the engaged intern. The so-called "secrets of success" are exposed when things work out well and approaches to avoid are evident when outcomes are less than desirable. Active and progressive involvement in projects, as well as in non-project activities, also helps to bond the intern to your organization.

The BOK explicitly requires experiential learning within 15 specific outcomes, as indicated by the outcomes in Figure 3 that include one or more E cells. The BOK requires practical experience to provide the context for these specific aspects of cognitive development. Experiential learning should be viewed as an extension of the cognitive development begun in universities. The 15 outcomes require your engagement with the engineer intern and the support of your organization.

Public and private sector engineering organizations are encouraged to engage in a broad approach to experiential learning by encouraging the temporary transfer of professionals in two directions. Practitioners can serve as adjunct university faculty and regular faculty can use sabbaticals and summer leave to work in engineering organizations. This two-way transfer will help to infuse the BOK across the profession.

### *Stress Client and Stakeholder Focus*

*Help the engineer intern understand the importance of learning client and stakeholder technical and nontechnical wants and needs.*

You know that many clients, as well as other users of engineering services, believe that they can easily find an engineer to solve a technical problem. Although the technical component is critical, advise the engineer intern that these clients and stakeholders want more than technical solutions. They seek engineers who can understand and identify with their environment and its unique set of issues, problems, opportunities, and constraints. If engineers, or more specifically, the intern's organization—whether it be an engineering firm or a public entity—does not fill these broad and deep wants and needs, other organizations will. Accordingly, urge engineer interns to further develop knowledge, skills, and attitudes found in such outcomes as outcome 8 (problem recognition and solving), outcome 11 (contemporary issues and historical perspectives), outcome 16 (communication—which, incidentally, includes asking questions to understand wants and needs), outcome 20 (leadership), and outcome 22 (attitudes).

Some clients and stakeholders, in their frustration with dealing with complex issues, take a position when initially interacting with engineers and other service providers that can be characterized by this statement: "I don't care how much you know until I know how much you care." You can urge the intern to help with the caring process, which includes doing the necessary "homework" about the client or stakeholder and asking many, wide-ranging questions. A problem well defined is half solved.

### *Support Licensure*

You are aware of evolving licensure requirements and the rationale behind them. Share this information with engineer interns. Explain how licensure benefits the individual, the organization, and the public. Remind the intern that licensure is one of the reasons for completing fulfillment of the BOK.

## Encourage Active Professional Society and Community Involvement

*Help the engineer intern see the personal and other benefits of active involvement in professional societies and local communities.*

Advise the engineer intern to immediately become actively, as opposed to passively, involved in at least one professional society such as ASCE. This is an effective way to continue one's personal and professional development in areas encompassed by outcome 16 (communication), outcome 20 (leadership), outcome 21 (teamwork), outcome 22 (attitudes), and outcome 23 (lifelong learning). You could explain to the intern that active professional society involvement supports the earlier advice of stressing client and stakeholder focus by providing a truer perspective of the real issues, challenges, and opportunities.

Furthermore, many civil engineers, including the engineer intern, derive a satisfying and prosperous living from their profession and, accordingly, should give something back to it. Explain that practicing engineers use the work of many predecessor professionals, most of whom produced the books, papers, conference proceedings, manuals of practice, computer software, and other valuable contributions for little or no monetary compensation.

Urge the engineer intern to also consider active participation in community organizations. Suggest that volunteer efforts enable many neighborhood, religious, and community-wide organizations to carry out useful functions. Given the progress that the intern has made toward fulfilling the broad and deep BOK, he or she is in an excellent position to begin to contribute to community activities. Examples are participating in an American Cancer Society fund-raiser, serving on an appointed community committee or board, assisting with the fund drives of a religious organization, coaching Special Olympics athletes, and running for elective office. Besides the value of the service that is provided, benefits to the intern are similar to those already noted for active participation in professional societies.

## Exemplify Professional Behavior

*Exemplify the personal and professional behavior that you extol.*

Most engineer interns will listen respectfully to advice offered by experienced practitioners. However, some interns may be skeptical as they look beyond your words. Your effectiveness as a coach or mentor will be enhanced if you serve as a positive role model and exemplify the personal and professional behavior that you extol. You can, by your actions, show the engineer intern the value of continued professional development, experiential learning, client and stakeholder focus, licensure, and active involvement in professional societies and community groups.

## Summary

Gradual implementation of the civil engineering BOK requires active participation by faculty members, students, engineer interns, and practitioners. Accordingly, this chapter offers guidance to members of each of the four groups in the hope that interested individuals will find insight and advice applicable to their particular situation.

# CHAPTER 5

# The Next Steps

*Do not follow where the path may lead.*
*Go instead where there is no path and leave a trail.*
Anonymous

*The BOK2 Committee has carried out its charge by refining the original BOK, incorporating Bloom's Taxonomy, addressing the roles of education and experience, and offering BOK-related guidance to stakeholders.*

With the completion of this, the second edition of *Civil Engineering Body of Knowledge for the 21st Century: Preparing the Civil Engineer for the Future,* the BOK2 Committee, which began its work in October 2005, has essentially completed its charge by publishing this report, which:

- Refines the original BOK to provide a clearer more usable description of the knowledge, skills, and attitudes needed to enter the practice of civil engineering at the professional level in the 21st century;
- Uses Bloom's Taxonomy, with its emphasis on action verbs, to describe levels of achievement for outcomes with the result being the BOK Outcome Rubric (Appendix I), and the non-prescriptive explanations for outcomes (Appendix J), which are the heart of this report,
- Addresses the role of the bachelor's degree, the master's degree or approximately 30 credits, and prelicensure experience in fulfilling the BOK and outlines the ways in which such fulfillment can be validated; and
- Offers guidance to those individuals who will play critical roles in using the BOK to implement PS 465—namely, faculty members, students, engineer interns, and practitioners.

In keeping with the charge, the process used by the BOK2 Committee to provide the preceding was transparent, inclusive, and interactive. The committee sought and welcomed questions and suggestions by corresponding members and other stakeholders. During the two-year course of the committee's work, individual members spoke and interacted with various groups.

The first edition of the BOK report, published in January 2004, served the civil engineering profession for four years. With its publication in early 2008, the second edition, which benefited from and is a substantial improvement over the first edition, should serve the profession for many years. Like the first edition, the second edition will stimulate curricula review, refinement, and design; encourage accreditation criteria advances; offer guidance for the education and training programs of private and public organizations that employ civil engineers; and support changes in licensure requirements. This latest edition, like the original, will also facilitate BOK and related discussions within engineering disciplines and societies within and outside of the U.S.

The BOK2 Committee believes that this report will significantly assist with further implementation of ASCE's master plan for implementing PS 465. The committee asks that stakeholders use the recommendations of this report as they move forward in carrying out their responsibilities. More specifically, the committee suggests the following for the indicated stakeholders:

*Stakeholders in the civil engineering community are asked to study and, to the extent feasible, use this report as they carry out their responsibilities.*

**ASCE CAP$^3$ Accreditation Committee:** Use this report as the basis of continued review of the Program Criteria for Civil and Similarly Named Engineering Programs, General Criteria for Masters Level Programs, the ASCE commentary, and the *Frequently Asked Questions* document. Coordinate with ASCE's representatives on the EAC to promote the EAC's support for further improvement to accreditation criteria.

**ASCE CAP$^3$ Licensure Committee:** Use this report to communicate to NCEES and licensing boards the levels of achievement recommended to practice civil engineering at the professional level. This is important in facilitating adoption of the NCEES Model Law, especially its increased education requirements, by each of the 55 U.S. licensing jurisdictions.[21]

**ASCE CAP$^3$ BOK Educational Fulfillment Committee:** Use this report to review the BOK and to further foster the creation of a community of scholars interested in educational reform and to document how programs could incorporate the BOK into their curricula.

**ASCE CAP$^3$ BOK Experiential Fulfillment Committee:** Use this report to prepare guidelines to assist the engineer intern in achieving those outcomes identified for partial fulfillment through on-the-job education and training.

**ABET, Inc.:** Use this report as a basis for adopting Bloom's Taxonomy as a common method for stating program outcomes within the various accreditation criteria.

**Departments of Civil and Environmental Engineering:** Consider using the civil engineering BOK and levels of achievement to be fulfilled via formal education when evaluating and designing bachelor's and master's degree programs. Civil and environmental engineering departments are also urged to give thought to using the affective domain of Bloom's Taxonomy.

**Employers of Civil Engineers:** Consider using the civil engineering BOK as one input when creating or revising professional development programs whose participants will include engineer interns. The BOK—especially the additional levels of achievement to be fulfilled during prelicensure experience—can help guide the content and conduct of seminars, workshops, mentoring, tutoring, coaching, experiential learning, and periodic personnel reviews. Encourage engineer interns to move toward licensure and stress the need to fulfill the BOK as a prerequisite for licensure.

**Civil Engineering Students and Interns:** Study the BOK, especially the 24 outcomes, and notice the portion to be fulfilled through formal education and the portion to be fulfilled via prelicensure experience. View the BOK as the road map by which you can travel toward your destination—entering the practice of civil engineering at the professional level—and measure your progress.

**Other Engineering Disciplines and Organizations:** To the extent many of us share interest in bodies of knowledge and in defining related achievement levels, CAP$^3$ would welcome comments on this report's findings and recommendations. Input received will be shared among the various groups working to implement ASCE PS 465.

# Acknowledgments

*If you want to go fast, go alone.*
*If you want to go far, go together.*
African proverb

The BOK2 Committee built on the work of others. The committee is indebted to individuals, committees, and other entities, within and outside of ASCE, including the BOK1 Committee, all of whom have and/or are contributing to the implementation of ASCE PS 465 and related reform initiatives. A special thank you to corresponding members of the BOK2 Committee and other contributors, all of whom are listed in Appendix E.

Equally important, the BOK2 Committee gratefully acknowledges the pioneering efforts of those individuals, committees, and other groups who, over the past several decades, advocated reform in the education and prelicensure experience of civil engineers. The breadth, depth, and influence of earlier initiatives is evident in the sources cited in various reports, such as this one, that have been or are being prepared as part of the master plan for implementation of ASCE PS 465. The BOK2 Committee believes that the earlier work is the root of what is being accomplished today and, as such, is bearing fruit.

*The findings and recommendations presented in this report build on the efforts of farsighted individuals, committees, and other groups dedicated to reform in the preparation of civil engineers.*

# APPENDIX A

# Abbreviations

| | |
|---|---|
| AACU | Association of American Colleges and Universities |
| AAES | American Association of Engineering Societies |
| ABET | Formerly Accreditation Board for Engineering and Technology (now simply ABET, Inc.) |
| AIChE | American Institute of Chemical Engineers |
| AOE | Approved Outside Entity |
| APWA | American Public Works Association |
| ASCE | American Society of Civil Engineers |
| ASEE | American Society for Engineering Education |
| ASME | American Society of Mechanical Engineers |
| B | Portion of the BOK fulfilled through the bachelor's degree |
| B+M/30&E | Bachelor's plus master's, or approximately 30 semester credits of acceptable graduate-level or upper-level undergraduate courses in a specialized technical area and/or professional practice area related to civil engineering, and experience. |
| B.S. | Bachelor of Science |
| BOK | Body of knowledge—that is, "the necessary depth and breadth of knowledge, skills, and attitudes required of an individual entering the practice of civil engineering at the professional level in the 21st Century."[2] |
| BOK1 | Body of knowledge as presented in the first edition of the ASCE BOK report[3] |
| BOK2 | Body of knowledge as presented in the second edition of the ASCE BOK report |

| | |
|---|---|
| BOKEdFC | BOK Educational Fulfillment Committee |
| BOKExFC | BOK Experiential Fulfillment Committee |
| BSCE | Bachelor of Science in Civil Engineering |
| CADD | Computer-aided drafting and design |
| CAP[3] | Committee on Academic Prerequisites for Professional Practice |
| CDIO | Conceive-Design-Implement-Operate |
| E | Portion of the BOK fulfilled through prelicensure experience |
| EAC | Engineering Accreditation Commission (of ABET) |
| EC | Engineering Criteria |
| ELQTF | Engineering Licensure Qualifications Task Force (of NCEES) |
| ES | Executive Summary |
| ETW | ExCEEd Teaching Workshop |
| ExCEEd | Excellence in Civil Engineering Education |
| FPD | First Professional Degree |
| H&SS | Humanities and social sciences |
| IEEE | Institute of Electrical and Electronics Engineers |
| IMF | International Monetary Fund |
| J.D. | Juris Doctor |
| KSA | Knowledge, skills, and attitudes |
| LOA | Levels of achievement as in reference to the CAP[3] Levels of Achievement Subcommittee and its report[9] |
| LQOC | Licensure Qualifications Oversight Group (of NCEES) |
| M | Formal post-baccalaureate education program that leads to a master's degree and to fulfillment of a portion of the requisite BOK |
| M.B.A. | Master of Business Administration |
| M.D. | Doctor of Medicine |
| MOE | Master's or equivalent |

| | |
|---|---|
| M/30 | Portion of the BOK fulfilled through the master's degree or equivalent (approximately 30 semester credits of acceptable graduate-level or upper-level undergraduate courses in a specialized technical area and/or professional practice area related to civil engineering) |
| M.S. | Master of science |
| NAE | National Academy of Engineering |
| NCEES | National Council of Examiners for Engineering and Surveying |
| NSF | National Science Foundation |
| NSPE | National Society of Professional Engineers |
| P.E. | Professional Engineer |
| PS | Policy statement |
| P&S | Probability and statistics |
| SAME | Society of American Military Engineers |
| TCAP[3] | Task Committee on Academic Prerequisites for Professional Practice |
| TCFPD | Task Committee for the First Professional Degree |
| UNESCO | United Nations Educational, Scientific, and Cultural Organization |
| UP&LG | Uniform Policies & Legislative Guidelines (an NCEES committee) |
| WTO | World Trade Organization |
| 30 | Approximately 30 semester credits of acceptable graduate-level or upper-level undergraduate courses in a specialized technical area and/or professional practice area related to civil engineering that does not lead to a formal master's degree but leads to the fulfillment of a portion of the requisite BOK |

# APPENDIX B

# Glossary

**Affective domain of learning:** "…of, or arising from, affects or feelings; emotional."[37] (See also the cognitive and psychomotor domains.)

**Attitudes:** The ways in which one thinks and feels in response to a fact or situation. Attitudes reflect an individual's values and world view and the way he or she perceives, interprets, and approaches surroundings and situations.

**Body of knowledge (BOK):** "…the necessary depth and breadth of knowledge, skills, and attitudes required of an individual entering the practice of civil engineering at the professional level in the 21st century."[8]

**Civil engineering:** "…the profession in which a knowledge of the mathematical and physical sciences gained by study, experience, and practice is applied with judgment to develop ways to utilize, economically, the materials and forces of nature for the progressive well-being of humanity in creating, improving and protecting the environment, in providing facilities for community living, industry and transportation, and in providing structures for the use of humanity."[6]

**Cognitive domain of learning:** "…of, or arising from, perception, memory and judgment."[37] (See also the affective and psychomotor domains of learning.)

**Critical thinking:** "…the intellectually disciplined process of actively and skillfully conceptualizing, applying, analyzing, synthesizing, and/or evaluating information gathered from, or generated by, observation, experience, reflection, reasoning, or communication, as a guide to belief and action. In its exemplary form, it is based on universal intellectual values that transcend subject matter divisions: clarity, accuracy, precision, consistency, relevance, sound evidence, good reasons, depth, breadth, and fairness."[38] "…skillful, responsible thinking that facilitates good judgment because it: 1) relies upon criteria; 2) is self-correcting; and 3) is sensitive to context."[39]

**Discovery learning:** "…type of learning whereby learners construct their own knowledge by experimenting with a domain, and inferring rules from the results of these experiments. The basic idea of this kind of learning is that because learners can design their own experiments in the domain and infer the rules of the domain themselves, they are actually constructing their knowledge. Because of these constructive activities, it is assumed they will understand the domain at a higher level than when the necessary information is just presented by a teacher or an expository learning environment."[40]

**Emerging technology:** A technical area of study and/or application that is based on a new material, test method, or design issue.

An emerging technology typically requires new design approaches, techniques to determine a specific engineering property or properties, or investigation tools. An emerging technology can originate from one or more traditional technologies, a new area of public concern, and/or public desire for improved infrastructure solutions. For example, the field of geosynthetics rapidly emerged from the civil, geotechnical, and environmental engineering technologies to perform a specific function of waste containment, based primarily on the public concern for this environmental issue.

**Humanities:** Includes disciplines that study the human condition—for example, philosophy, history, literature, the visual arts, the performing arts, language, and religion.

**Knowledge:** Is largely cognitive and consists of theories, principles, and fundamentals. Examples are geometry, calculus, vectors, momentum, friction, stress and strain, fluid mechanics, energy, continuity, and variability.

**Outcome:** Statements that describe what individuals are expected to know and be able to do by the time of entry into the practice of civil engineering at the professional level in the 21st century. Outcomes define the knowledge, skills, and attitudes that individuals acquire through appropriate formal education and prelicensure experience.

**Practice of civil engineering at the professional level:** "Practice as a licensed professional engineer."[2]

**Psychomotor domain of learning:** "...of, or arising from, the motor effects of mental processes."[37] (See also the affective and cognitive domains of learning.)

**Rubric:** A set of instructions or an explanation; something under which a thing is classed.[14]

**Skill:** The ability to perform tasks. Examples are using a spreadsheet; continuous learning; problem solving; critical, global, integrative/system, and creative thinking; teamwork; communication; and self-assessment.

**Social Sciences:** Includes disciplines that study the human aspects of the world—for example, economics, political science, sociology, psychology, and anthropology.

**Sustainability:** The ability to meet human needs for natural resources, industrial products, energy, food, transportation, shelter, and effective waste management while conserving and protecting environmental quality and the natural resource base essential for the future.[41]

**Sustainable development:** "...the challenge of meeting human needs for natural resources, industrial products, energy, food, transportation, shelter, and effective waste management while conserving and protecting environmental quality and the natural resource base essential for future development."[42,43]

**Sustainable engineering:** Meeting human needs for natural resources, industrial products, energy, food, transportation, shelter, and effective waste management while conserving and protecting environmental quality and the natural resource base essential for future development.[44]

**Systems analysis:** The formulation and exercise of a model to answer a question or address a problem concerning a system.

**Team—intradisciplinary:** Consists of members from within the civil engineering subdiscipline—for example, a structural engineer working with a geotechnical engineer.

**Team—multidisciplinary:** Composed of members from different professions—for example, a civil engineer working with an economist. Multidisciplinary also includes a team consisting of members from different engineering subdisciplines (sometimes referred to as a cross-disciplinary team).

# APPENDIX C

# ASCE Policy 465: Emergence of the Body of Knowledge

In October 1998, following years of studies and conferences, the ASCE Board of Direction adopted Policy Statement 465 (PS 465), which began as follows: "The ASCE supports the concept of the master's degree as the first professional degree (FPD) for the practice of civil engineering at the professional level." Partly as a result of the October 2001 report[45] of the board's Task Committee for the First Professional Degree (TCFPD), the board adopted a revised PS 465 in 2001 titled "Academic Prerequisites for Licensure and Professional Practice." The revised policy said: "ASCE supports the concept of the master's degree or equivalent (MOE) as a prerequisite for licensure and the practice of civil engineering at the professional level."

The ASCE board created the Task Committee on Academic Prerequisites for Professional Practice (TCAP³) in October 2001 to build on the work of the TCFPD. TCAP³ was charged to "...develop, organize, and execute a detailed plan for full realization of ASCE PS 465." With the formation of TCAP³, PS 465 was moving from the study phase to the implementation phase. TCAP³ became the Committee on Academic Prerequisites for Professional Practice (CAP³) in 2003.

In response to a CAP³ recommendation, the ASCE board revised PS 465 in October 2004 so as to replace the MOE language with the body of knowledge (BOK).[8] The policy now reads, in part:

> *The ASCE supports the attainment of a body of knowledge for entry into the practice of civil engineering at the professional level. This would be accomplished through the adoption of appropriate engineering education and experience requirements as a prerequisite for licensure.*

ASCE PS 465 was refined by the ASCE board for a fourth time in April 2007.[8] While the changes were relatively minor, the policy clearly states that the BOK includes: 1) fundamentals of mathematics, science, and engineering science; 2) technical breadth; 3) breadth in the humanities and social sciences; 4) professional practice breadth; and 5) technical depth or specialization. This is consistent with *The Vision for Civil Engineering in 2025*[1] and the model for educational preparation for civil engineering practice (see Figure K-1).

The BOK is defined in the policy as "the necessary depth and breadth of knowledge, skills, and attitudes required of an individual entering the practice of civil engineering at the professional level in the 21st century."

The long-term effect of PS 465 is illustrated in Figure C-1, which compares today's civil engineering professional track

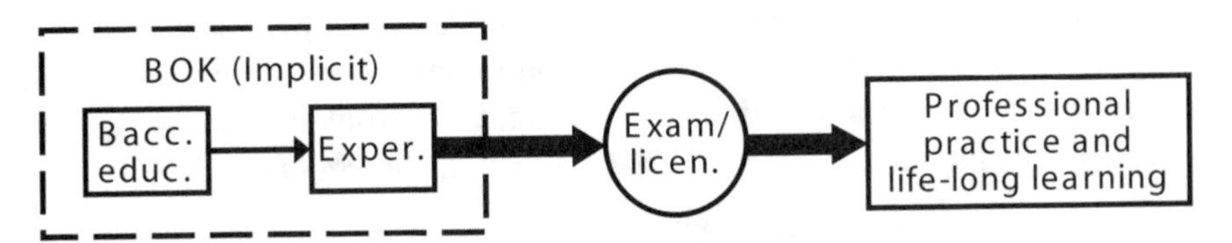


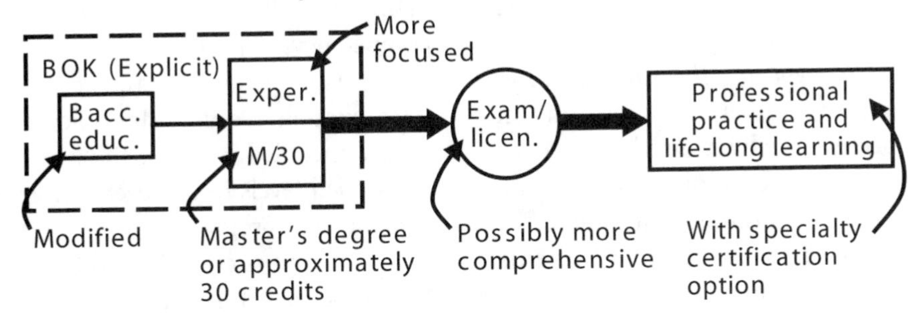


Figure C-1. Implementation of Policy Statement 465 will improve the lifelong career of tomorrow's civil engineer.

with tomorrow's. Intended changes include:

- an explicit BOK,
- a redesigned baccalaureate program,
- a master's degree or approximately 30 semester credits of acceptable graduate-level or upper-level undergraduate courses in a specialized technical area and/or professional practice area related to civil engineering,
- a more focused prelicensure experience,
- a more comprehensive licensure examination, and
- the option of specialty certification.

From ASCE's perspective, the BOK represents a strategic direction for the profession. Under today's accreditation and regulatory processes and procedures, some of the elements of the BOK may not be translated into curricula, accreditation criteria, and licensing requirements in the near term. In other words, the BOK describes the "gold standard" for the aspiring civil engineering professional. Because input into curricula design, accreditation, and licensing comes from many and varied stakeholders beyond ASCE, these processes are not likely to reflect all aspects of ASCE's BOK. ASCE is optimistic that the curricula design, accreditation, and licensing processes will change over time to adopt a more BOK-centric approach. As this occurs, a greater proportion of the BOK will be reflected in curricula and in accreditation and licensure requirements.

The preceding brief history of ASCE PS 465 reveals two essentials.

- The ASCE Board of Direction has been consistent in its 1998 initial adoption and subsequent 2001, 2004, and 2007 refinements of the policy. ASCE leadership strongly supports reform of civil engineering education and prelicensure experience.
- The premise of PS 465 gradually shifted from a degree basis (for example, "the master's as the first professional

degree") through the "master's degree or equivalent" approach, finally settling on a BOK foundation. This provides flexibility for engineers who cannot or do not wish to pursue a master's degree through traditional means.

The BOK thrust resulted in the CAP$^3$ Body of Knowledge Committee completing, in January 2004, the report *Civil Engineering Body of Knowledge for the 21st Century: Preparing the Civil Engineer for the Future.*[3] Deliberations resulting in that report eventually viewed reform of the process by which individuals prepare for entry into the professional practice of civil engineering as having three elements or standing on three legs. They are: 1) ***what*** should be taught and learned; 2) ***how*** should it be taught and learned; and 3) ***who*** should teach and learn it. The committee's primary focus was the ***what.***

The ***what*** recommendations were cast in terms of 15 outcomes that, compared to today's bachelor's programs, included significant increases in technical and professional depth. Included in the 15 outcomes were the 11 outcomes similar to those used by ABET. Each outcome was further described with a civil engineering commentary.

As a result of reviewing and using the recommendations in the civil engineering BOK report, stakeholders identified a problem and raised issues related to the BOK. The problem revolved around the three principal words used to define competency levels—namely, recognition, understanding, and ability. In particular, the CAP$^3$ Curriculum Design Committee came to this conclusion: Until there were understandable and readily applicable competency definitions—including definitions that would be understood by those outside of the committee—development of model curricula would be fruitless because they may not achieve the intent of the BOK.

To remove this obstacle, CAP$^3$ formed the Levels of Achievement Subcommittee in February 2004 to resolve the levels of competency problem. Subcommittee membership included the chair of the ABET Accreditation Council Task Force, whose charge included studying the inconsistency in implied levels of achievement across the general and program criteria of the four ABET commissions. The subcommittee's September 2004 report[9] contained many recommendations that solved the problem and are being implemented. Relative to this second edition BOK report, the Subcommittee recommended:

- Substituting "achievement" for competency in all future references to levels of demonstrated learning.
- Asking the CAP$^3$ Accreditation Committee to use the revised outcomes as the basis for drafting Program Criteria for Civil and Similarly Named Engineering Programs and General Criteria for Masters-Level Programs.
- Using Bloom's Taxonomy to define levels of achievement. Bloom's levels of the cognitive domain are widely known and understood within the education community. Furthermore, use of measurable, action-oriented verbs facilitates more consistent curricula design and assessment.

# APPENDIX D

# Charge to the Body of Knowledge Committee

## Introduction

The first edition of *Civil Engineering Body of Knowledge for the 21st Century: Preparing the Civil Engineer for the Future,* which was released in January 2004, was well received by a cross section of the U.S. civil engineering community and even beyond the U.S. The body of knowledge (BOK) structure is proving to be a productive common ground for discussion among civil engineering academics and practitioners as well as members of other engineering disciplines.

The first edition of the BOK was envisioned as a work-in-progress that would be updated on the basis of input from stakeholders within and outside of civil engineering. The first edition has generated significant discussion that has, in turn, produced helpful questions, critiques, and suggestions. Accordingly, the ASCE Committee on Academic Prerequisites for Professional Practice (CAP$^3$), the group charged with implementing ASCE PS 465, will form the Second Edition of the Body of Knowledge Committee. This committee will be charged, as detailed below, with producing an improved second edition of the BOK report in response to recent stakeholder input and other developments in engineering education and practice.

## Charge

The Second Edition Body of Knowledge Committee is asked to:

1. Collect and review stakeholder input received since publication of the first edition.
2. Help to publicize the committee's work with the goal of seeking additional input from a broad community of stakeholders.
3. Objectively assess the substance of the civil engineering BOK, as presented in the first edition. Identify issues, beyond those listed in this charge that may require attention.
4. Review the findings and recommendations of the Attitudes Study Committee, a subcommittee of the CAP$^3$ Curriculum Committee, and reflect them in the second edition.
5. Review the findings and recommendations of the CAP$^3$ Levels of Achievement Subcommittee and revise the BOK accordingly. In particular, follow through on the following with respect to the subcommittee's report:
   - Revise the statements of the outcomes using verbs based on Bloom's Taxonomy.

- Refine and use the BOK outcomes rubrics table.
- Respond to the discussion of critical thinking.

6. Strengthen and/or more explicitly discuss the humanities and social sciences (some refer to this as liberal or general education) content of the BOK.
7. Replace the word "commentary," as used in the first edition, with an appropriate word such as "explanation." Reason: Eliminate possible confusion with the use of "commentary" in explanations of ABET criteria.
8. Examine the findings and recommendations of the National Academy of Engineering reports, *The Engineer of 2020: Visions of Engineering in the New Century* and *Educating the Engineer of 2020: Adapting Engineering Education to the New Century*, which were published after the first edition of the BOK. Describe how the two NAE reports support and/or differ from the civil engineering BOK and respond accordingly.
9. Communicate with CAP[3] and its accreditation, curriculum design, and licensure committees and other special purpose groups it may form.
10. Communicate with such groups within ASCE having BOK interests as the Educational Activities Committee and its constituent committees; the American Academy of Water Resource Engineers; Civil Engineering Certification, Inc.; and the Civil Engineering Department Heads Council and its executive committee.
11. Communicate with such groups outside of ASCE as ABET's Research and Assessment office, the ABET Accreditation Council Task Force or its successor, the National Council of Examiners for Engineering and Surveying, the National Society of Professional Engineers, the American Society of Mechanical Engineering, the Institute for Electronics and Electrical Engineers, the U.S. Army Corps of Engineers, the National Academy of Engineering, the Society of American Military Engineers, and other organizations having an interest in the BOK.
12. Prepare a draft report, for review by CAP[3], by October 15, 2006, to allow for publication in January 2007. (Note: The full final BOK2 report was published about one year later than expected. Reasons for the extra time included the unexpected complexity of drafting, discussing, and concurring on the rubric and explanations and a major effort to seek input from stakeholders.) That report should:
    - Provide a record of the committee's process, actions, findings, and recommendations.
    - Follow the overall structure of the first edition but reduce the length. The reduced length can be accomplished by omitting certain supplementary materials that are in the first edition and can be referenced. The second edition should also move as much material as possible from the report's body into its appendices. (Note: The second edition is larger than the first edition because of complexities encountered, research conducted, and the decision to document research results in appendices. The goal of a shorter report will be achieved by the decision to publish and widely distribute a very short summary report that focuses on the rubric, which is the heart of the full report.)

- Retain the "what," "how," and "who" dimensions of the BOK, as was done in the first edition, but continue to place the primary emphasis on the "what."

13. Publish and widely disseminate the report. Upon approval of the committee's report by CAP[3], ASCE will be asked to provide editing services and to print and distribute the report. The second edition will be similar in "look" and quality to the first edition but will use cover color/graphics to clearly distinguish it.

## Committee Composition

CAP[3] envisions a second edition committee composed of from 8 to 15 individuals and at least one individual representing each of the following stakeholder groups or entities:

1. Member of the executive committee of the ASCE Civil Engineering Department Heads Council.
2. Member(s) of the civil engineering educational community.
3. Member(s) of the private civil engineering practice community.
4. Member(s) of the public civil engineering practice community.
5. A current civil engineering student or recent civil engineering graduate.
6. Representative from the ABET staff knowledgeable of outcome formulation and assessment.
7. Member of the CAP[3] Accreditation Committees (could be a liaison person instead of a committee member).
8. Member of the CAP[3] Curriculum Design Committee (could be a liaison person instead of a committee member).
9. Member of the CAP[3] Licensure Committee (could be a liaison person instead of a committee member).
10. Member of ASCE Civil Engineering Certification, Inc. (could be a liaison person instead of a board member).

## Effort Expected of Committee Members

1. Commit to active involvement throughout the expected 18-month life of the committee (from about October 2005 to April 1, 2007).
2. Participate in about two face-to-face meetings, which will be held in a cost-effective location and occur all day Saturday and half of Sunday. Most expenses will be reimbursed in accordance with ASCE policy.
3. Participate in one-hour conference calls to be held every two to three weeks.
4. Equitably volunteer for research, writing, and presentation tasks and/or accept task assignments as needed to carry out the committee's charge.

Prepared by CAP[3]
September 2005

# APPENDIX E

# Members and Corresponding Members of the Body of Knowledge Committee

## Members

CAP$^3$ sought a new and diverse group of engineering practitioners and educators as members and corresponding members of the new BOK2 Committee. Of particular interest to CAP$^3$ were potential members who could provide a fresh assessment of the BOK.

To accomplish this, CAP$^3$ took a unique approach in recruiting the members of the new committee. In the August 2005 issue of *ASCE News*, a public call was made to seek self-nominations for the new committee. Each nominee completed a comprehensive application describing their background, interest, and commitment to the BOK project.

Approximately 30 applications were received from this open call. Selections of BOK2 Committee members and a core group of corresponding members were made by CAP$^3$ in September 2005. In October 2005, an ASCE PS 465 workshop was held for new committee members in conjunction with the ASCE annual meeting in Los Angeles. This one-day workshop addressed the background of PS 465 and the BOK—and ASCE's progress in implementation. Following this workshop the new members were asked to confirm their interest in the BOK2 Committee. With a core group of 14 members and more than 20 initial corresponding members, the BOK Committee had its first weekly telephone conference on November 16, 2005, and its first face-to-face meeting in Tampa, Florida, on January 28–29, 2006.

Members of the BOK2 Committee are:

Richard O. ANDERSON, P.E., Hon.M.ASCE, Somat Engineering, Detroit, MI, roape1@aol.com. (Chairperson)

Kenneth J. FRIDLEY, Ph.D., M.ASCE, Department of Civil and Environmental Engineering, University of Alabama, Tuscaloosa, AL, kfridley@eng.ua.edu. (Vice Chair)

Stuart G. WALESH, Ph.D., P.E., Hon.M.ASCE, Consultant, Englewood, FL, stuwalesh@comcast.net. (Editor)

Anirban DE, Ph.D., P.E., M.ASCE, Department of Civil Engineering, Manhattan College, Riverdale, NY, anirban.de@manhattan.edu.

Decker B. HAINS, Ph.D., P.E., M.ASCE, U.S. Military Academy, West Point, NY, decker.hains@us.army.mil.

Ronald S. HARICHANDRAN, Ph.D., P.E., F.ASCE, Department of Civil and Environmental Engineering, Michigan State University, East Lansing, MI, harichan@egr.msu.edu.

Peter W. HOADLEY, Ph.D., P.E., A.M.ASCE, Department of Civil Engineering, Virginia Military Institute, Lexington, VA, hoadley@vmi.edu.

Manoj K. JHA, Ph.D., P.E., M.ASCE, Department of Civil Engineering, Morgan State University, Baltimore, MD, mkjha@eng.morgan.edu.

David A. LANGE, Ph.D., P.E., M.ASCE, Department of Civil and Environmental Engineering, University of Illinois at Urbana-Champaign, IL, dlange@uiuc.edu.

Melanie L. LAWRENCE, A.M.ASCE, Leonard Rice Engineers, Inc., Denver, CO, lawrence@lrcwe.com.

Timothy F. LENGYEL, P.E., M.ASCE, Winzler & Kelly Consulting Engineers, San Francisco, CA, timlengyel@w-and-k.com.

Daniel R. LYNCH, Ph.D., M.ASCE, Thayer School of Engineering, Dartmouth College, Hanover, NH, daniel.r.lynch@dartmouth.edu.

Robert E. MACKEY, P.E., M.ASCE, S2L Inc., Maitland, FL, bmackey@s2li.com.

John M. MASON, Ph.D., P.E., M.ASCE, College of Engineering, Pennsylvania State University, University Park, PA, jmm7@psu.edu.

Jeffrey S. RUSSELL, Ph.D., P.E., F.ASCE, Department of Civil and Environmental Engineering, University of Wisconsin-Madison, WI, russell@engr.wisc.edu. ($CAP^3$ Chair)

## Corresponding Members

The purpose of inviting individuals to participate as corresponding members of the BOK2 Committee was to further encourage an active, open dialogue on the topics of discussion relevant to members of the ASCE family and beyond. By doing so, the BOK2 Committee believed that even more rational, equitable decisions could be reached that would help implement ASCE PS 465—that is, reform of the education and prelicensure experience of civil engineers. As indicated by the following list, more than 50 individuals representing a wide variety of professional situations chose to participate as corresponding members.

These corresponding members were copied on essentially all draft and other materials distributed to the BOK2 Committee. They were also informed of planned conference calls and face-to-face meetings of the committee and invited to participate as interest and time permitted. Accordingly, corresponding members received meeting agendas and minutes. Finally, corresponding members were invited to participate in the many e-mail discussions, and sometimes debates, that occurred during the work of the BOK2 Committee, which they did with valuable insight and with vigor.

Carsten D. AHRENS, Ph.D., Fachhochschule Oldenburg, Oldenburg, Germany, carsten.ahrens@fh-oldenburg.de.

Alfredo H. S. ANG, Ph.D., Hon.M.ASCE, Bellevue, WA, ahang2@aol.com.

Tomasz ARCISZEWSKI, Ph.D., A.M.ASCE, Department of Civil, Environmental, and Infrastructure Engineering, George Mason University, Fairfax, VA, tarcisze@gmu.edu.

C. Robert BAILLOD, Ph.D., P.E., F.ASCE, DEE, Department of Civil and Environmental Engineering, Michigan Technological University, Houghton, MI, baillod@mtu.edu.

Amitabha BANDYOPADHYAY, Ph.D., P.E., M.ASCE, Holbrook, NY, bandyoa@farmingdale.edu.

Bryan BOULANGER, Ph.D., Zachry Department of Civil Engineering, Texas A&M University, College Station, TX.

Brian R. BRENNER, P.E., M.ASCE, Department of Civil and Environmental Engineering, Tufts University, Medford, MA, brian.brenner@tufts.edu.

Jason BURKE, P.E., M.ASCE, Big Timber, MT, jburkepe@aim.com.

Donald D. CARPENTER, A.M.ASCE, Lawrence Technological University, Southfield, MI, carpenter@ltu.edu.

Pascale CHAMPAGNE, Ph.D., A.M.ASCE, Department of Civil Engineering, Queen's University, Kingston, ON, Canada, champagne@civil.queensu.ca.

Karen C. CHOU, Ph.D., P.E., F.ASCE, Department of Mechanical and Civil Engineering, Minnesota State University, Mankato, MN, karen.chou@mnsu.edu.

Larry A. ESVELT, Ph.D., P.E., M.ASCE, Esvelt Environmental Engineering, Spokane, WA, larry@esvelt.com.

Robert ETTEMA, M.ASCE, Institute of Hydraulic Research, University of Iowa, Iowa City, IA, robert-ettema@uiowa.edu.

Jeffrey C. EVANS, Ph.D., P.E., M.ASCE, Department of Civil and Environmental Engineering, Bucknell University, Lewisburg, PA, evans@bucknell.edu.

Howard C. GIBBS, P.E., M.ASCE, Potomac Electric Power Company, Washington, DC, hcgibbs@pepco.com.

Ali HAGHANI, Ph.D., M.ASCE, Department of Civil Engineering, University of Maryland, College Park, MD, haghani@eng.umd.edu.

Gerd W. HARTUNG, P.E., M.ASCE, Bloomfield, MI, gwhartung@harleyellis.com.

Chris HENDRICKSON, Ph.D., M.ASCE, Department of Civil Engineering, Carnegie Mellon University, Pittsburgh, PA, cth@cmu.edu.

Thomas HEWETT, Department of Psychology, Drexel University, Philadelphia, PA, hewett@drexel.edu.

Garabed M. HOPLAMAZIAN, P.E., M.ASCE, Southfield, MI, garhoplamazian@akahn.com.

Kenneth C. JOHNS, Civil Engineering, Universite de Sherbrooke, Sherbrooke, QC, Canada, kenneth.johns@usherbrooke.ca.

Gudrun KAMMASCH, Ph.D., P.E., University of Applied Sciences, Berlin, Germany, kammasch@tfh-berlin.de.

Dinesh R. KATTI, Ph.D., P.E., M.ASCE, Department of Civil Engineering, North Dakota State University, Fargo, ND, dinesh.katti@ndsu.edu.

Kenneth G. KELLOGG, P.E., M.ASCE, Klamath Falls, OR, kenneth.kellogg@oit.edu.

William E. KELLY, P.E., F.ASCE, Catholic University of America, Washington, DC, kellyw@cua.edu.

Merlin KIRSCHENMAN, P.E., M.ASCE, Professor Emeritus, North Dakota State University, Moorhead, MN, m.kirschenman@ndsu.edu.

William R. KNOCKE, Ph.D., P.E., F.ASCE, Department of Civil Engineering, Virginia

Polytechnic Institute, Blacksburg, VA, knocke@vt.edu.

Kenneth W. LAMB, A.M.ASCE, G.C. Wallace, Las Vegas, NV, klamb@gcwallace.com.

Jim LAMMIE, P.E., Hon.M.ASCE, Consultant, Princeton, NY, lammie@pbworld.com.

William H. LEDER, P.E., F.ASCE, Houghton, MI, bleder@leaelliott.com.

James L. LEE, Historic American Engineering Record, Washington, DC, larry_lee@nps.gov.

E. Walter LEFEVRE, Ph.D., P.E., Hon.M.ASCE, Department of Civil Engineering, University of Arkansas, Fayetteville, AR, ewl@uark.edu.

Jerry J. MARLEY, Ph.D., P.E., M.ASCE, University of Notre Dame, Notre Dame, IN, marley.1@nd.edu.

Paul W. MCMULLIN, Ph.D., P.E., M.ASCE, Chief Engineer, Dunn Associates, Inc., Salt Lake City, UT, pmcmullin@dunn-sc.com.

Donald E. MILKS, Ph.D., P.E., F.ASCE, Chautauqua, NY, donace@netsync.net.

Adi K. MURTHY, P.A.K. Murthy Consultants, Chennai, India, murthyadi@hotmail.com.

James K. NELSON, Ph.D., P.E., F.ASCE, College of Engineering and Computer Science, University of Texas at Tyler, TX, jknelson@uttyler.edu.

John S. NELSON, P.E., Department of Civil and Environmental Engineering, University of Wisconsin-Madison, WI, napco57@tds.net.

James K. PLEMMONS, Ph.D., P.E., M.ASCE, The Citadel, Charleston, SC, keith.plemmons@citadel.edu.

Stephen J. RESSLER, Ph.D., P.E., Hon.M.ASCE, U.S. Military Academy, West Point, NY, stephen.ressler@usma.edu.

Jerry R. ROGERS, Ph.D., P.E., D.WRE, F.ASCE, Civil and Environmental Engineering Department, University of Houston, Houston, TX, jerryrogers@houston.rr.com.

David I. RUBY, P.E., F.ASCE, Ruby & Associates PC, Farmington Hills, MI, druby@rubyusa.com.

Steven D. SANDERS, P.E., M.ASCE, GSW & Associates Inc., Dallas, TX, ssanders@gsw-inc.com.

Subal SARKAR, P.E., M.ASCE, Princeton Junction, NJ, sarkars@pbworld.com.

David M. SCHWEGEL, A.M.ASCE, Sacramento, CA, dschwegel@stantec.com.

Roger K. SEALS, Ph.D., P.E., F.ASCE, Department of Civil and Environmental Engineering, Louisiana State University, Baton Rouge, LA, ceseal@lsu.edu.

Jennifer Walker SHANNON, P.E., M.ASCE, TBE Group, Inc., Clearwater, FL, jolivin@yahoo.com.

Alan T. SHEPPARD, P.E., M.ASCE, Strongsville, OH, atshepp@earthlink.net.

Johann F. SZAUTNER, P.E., L.S., M.ASCE, Bethlehem, PA, jfs@cowanassociates.com.

Y. C. TOKLU, Ph.D., P.E., M.ASCE, Faculty of Engineering, Bahcesehir University, Istanbul, yct001@gmail.com.

Marlee A. WALTON, P.E., M.ASCE, Iowa State University, Ames, IA, marlee@iastate.edu.

## ASCE Staff

Thomas A. LENOX, Ph.D., M.ASCE, tlenox@asce.org. (CAP$^3$ Staff Leader)

James J. O'BRIEN, Jr., P.E., M.ASCE, jobrien@asce.org. (Staff Contact)

Deborah CONNOR, dconnor@asce.org. (Staff Contact)

## Contributors to Special Tasks

The following individuals, although not necessarily BOK2 Committee members or corresponding members, kindly contributed to the indicated special tasks:

Anna M. MICHALAK, Ph.D., P.E., Department of Civil and Environmental Engineering and Department of Atmospheric, Oceanic, and Space Sciences, University of Michigan, served on the Risk/Uncertainty Task Group along with BOK2 Committee member Robert E. MACKEY and BOK2 Corresponding members Alfredo H. S. ANG and Karen C. CHOU.

Kevin G. SUTTERER, Ph.D., P.E., Department of Civil Engineering, Rose-Hulman Institute of Technology, along with BOK2 Committee member Daniel R. LYNCH and BOK2 Corresponding member Jeffrey C. EVANS, prepared Appendix G, The Affective Domain of Bloom's Taxonomy.

Ernest T. SMERDON, Ph.D., P.E.; Gerald E. GALLOWAY, Jr., Ph.D., P.E.; BOK2 Corresponding member Merlin KIRSCHENMAN; and BOK2 Committee member Kenneth J. FRIDLEY developed the Guidance for Faculty section of Chapter 4.

Debra LARSON, Ph.D., P.E., Professor and Chair, Department of Civil and Environmental Engineering, Northern Arizona University, and BOK2 Corresponding member Brian R. BRENNER contributed ideas and information used to create the Guidance for Students section of Chapter 4.

Bernard R. BERSON, P.E., L.S., P.P., FNSPE and NSPE President-Elect (2006–2007); Phillip E. BORROWMAN; P.E., Senior Vice President, Hanson Professional Services; and BOK2 Corresponding member Steven D. SANDERS contributed ideas and information used to create the Guidance for Engineer Interns section of Chapter 4.

William M. HAYDEN, Ph.D., P.E., Management Quality by Design; William S. HOWARD, P.E., Executive Vice President and Chief Technical Officer, CDM; and BOK2 Committee members Melanie L. LAWRENCE and Robert E. MACKEY provided ideas used to create the Guidance for Practitioners section of Chapter 4.

Richard L. CORRIGAN, Senior Vice President and Director of Strategic Initiatives, CH2M HILL, helped BOK2 Committee members John M. MASON and Timothy F. LENGYEL and BOK2 Corresponding members Steven D. SANDERS and Jennifer Walker SHANNON prepare Appendix N, Public Policy.

BOK2 Corresponding member Jeffrey C. EVANS and BOK2 Committee members Daniel R. LYNCH and David A. LANGE prepared Appendix K, Humanities and Social Sciences.

# APPENDIX F

# Bloom's Taxonomy

The articulation of BOK learning outcomes and related levels of achievement comes, in part, from the desire to clarify what should be taught and learned. Clarification can be achieved through the use of Bloom's Taxonomy of Educational Objectives[a] for the cognitive domain, which systematically differentiates outcome characteristics and promotes common understanding for all users of the BOK. The cognitive domain refers to educational objectives that involve the recall and recognition of knowledge and the development of intellectual abilities and skills.

Bloom's Taxonomy was originally conceived as a technique to reduce the labor of preparing comprehensive examinations through the exchange of test items among faculty at various universities.[a] The goal was to create banks of test items in which each bank attended to the same educational objective. A team of measurement specialists began meeting in 1949 to create the taxonomy of objectives and their first draft was published in 1956. Bloom believed, however, that the original taxonomy went beyond measurement. Among his many ideas was his belief that the taxonomy could serve as a common language for expressing and understanding learning goals or objectives.[b]

Bloom's emphasis on the use of measurable, action-oriented verbs facilitates the creation of outcome statements that lend themselves to more consistent and more effective assessment. Bloom's Taxonomy consists of six levels in the cognitive domain, which herein are called levels of achievement. These achievement levels for cognitive development will occur as a result of formal education and experience.

The *Levels of Achievement Subcommittee Report*[c] details the recommendation to use Bloom's Taxonomy as the levels of achievement for the BOK. The purpose of this appendix is to define the achievement levels and provide definitions of the active verbs used in the BOK for each level. These definitions are helpful because some of the active verbs can be used at different levels. Moreover, for some outcomes, Bloom's Taxonomy was not directly applicable and verbs were chosen with specific definitions to convey the progression through the levels of achievement. These special instances are noted in the definitions at each level. The definitions of the verbs were taken from *Webster's Third New International Dictionary, Unabridged.*[d] The definition of the levels of achievement were summarized from Bloom's Taxonomy of Educational Objectives,[a] *Stating Objectives for Classroom Instruction,* 2nd Edition[e] and from the Levels of Achievement Subcommittee Report.[c]

## Level 1—Knowledge

Knowledge is defined as the remembering of previously learned material. This may involve the recall of a wide range of material, from specific facts to complete theories, but all that is required is the bringing to mind of the appropriate information. Knowledge represents the lowest level of learning outcomes in the cognitive domain.[e]

*Define*: to discover and set forth the meaning of.

*Describe*: to present distinctly by means of properties and qualities.

*Identify*: to select; to choose something for a number or group.

*List*: to declare to be.

*Recognize*: to perceive clearly.

Other illustrative verbs at the *knowledge* level include: enumerate, label, match, name, reproduce, select, and state.

## Level 2—Comprehension

Comprehension is defined as the ability to grasp the meaning of material. This may be shown by translating material from one form to another (words to numbers), by interpreting material (explaining or summarizing), and by estimating future trends (predicting consequences or effects). These learning outcomes go one step beyond the simple remembering of material, and represent the lowest level of understanding.[e]

*Explain*: to make plain or understandable.

*Describe*: to present distinctly by means of properties and qualities.

*Distinguish*: to perceive as being separate or different.

*Discuss*: to present in detail.

Other illustrative verbs at the *comprehension* level include: classify, cite, convert, estimate, generalize, give examples, paraphrase, restate (in own words), and summarize.

## Level 3—Application

Application refers to the ability to use learned material in new and concrete situations. This may include the application of such things as rules, methods, concepts, principles, laws, and theories. Learning outcomes in this area require a higher level of understanding than those under comprehension.[e]

*Solve*: to find an answer, solution, explanation, or remedy for.

*Apply*: to use for a particular purpose or in a particular case.

*Use*: to carry out a purpose or action by means of.

*Formulate*: to plan out in orderly fashion.

*Develop*: to make clear, plain, or understandable. Develop is similar to "explain" but at a greater level of detail.

*Conduct*: the act, manner, or process of carrying out (as a task) or carrying forward.

*Report*: to give an account of; to give a formal or official account or statement of.

*Organize*: to put in a state of order.

*Function*: to carry on in a certain capacity.

*Demonstrate*: to illustrate or explain in an orderly and detailed way especially with many examples, specimens, and particulars.

*Explain:* to give the reason for or cause of. Although commonly a level 2 or level 5 verb when used in the context of outcome 11, contemporary issues and historical perspectives, the verb "explain" conveys the application of broad education to the identification, formulation, and solution of engineering problems.

Other illustrative verbs at the *application* level include: administer, articulate, calculate, chart, compute, contribute, establish, implement, prepare, provide, and relate.

## Level 4— Analysis

Analysis refers to the ability to break down material into its component parts so that its organizational structure may be understood. This may include the identification of parts, analysis of the relationship between parts, and recognition of the organizational principles involved. Learning outcomes here represent a higher intellectual level than comprehension and application because they require an understanding of both the content and the structural form of the material.[e]

*Analyze*: to ascertain the components of or separate into component parts; determine carefully the fundamental elements of (as by separation or isolation) for close scrutiny and examination of constituents or for accurate resolution of an overall structure or nature.

*Select*: to choose something from a number or group.

*Organize*: to arrange by systematic planning and coordination; to unify into a coordinated functioning whole. Although "organize" is not typically a level 4 verb, it is appropriate for outcomes 8 (problem recognition and solving), 16 (communication), and 20 (leadership). For each of these outcomes, the verb "organize" conveys the appropriate educational objective progression.

*Compare*: to examine the character or qualities of, especially for the purpose of discovering resemblances or differences.

*Contrast*: to compare in respect of differences; to examine like objects by means of which dissimilar qualities are made prominent.

*Illustrate*: to make clear by giving examples or instances.

*Formulate*: to put into a systematized statement or expression.

*Deliver*: give forth in words; to make known to another. Although the verb "deliver" is not typically a level 4 verb, it is appropriate for outcome 16 (communication) because it conveys the appropriate educational objective progression.

*Function*: to carry on in a certain capacity. For level 4, the verb "function" is only used for outcome 21 (teamwork) and it has the same definition at level 4 as it does at level 3. In this case, the verb does not convey the educational progression between levels 3 and 4. Rather, the progression is delineated by the movement from an intradisciplinary to a multidisciplinary team.

*Direct*: to carry out the organizing, energizing, and supervising of, especially in an authoritative capacity; to regulate the activities or course of; to guide and supervise; to assist by giving advice, instruction, and supervision. The verb "direct" may not be considered a typical level 4 verb; however, within the context of outcome 20 (leadership), the verb "direct" conveys the logical educational progression in the outcome.

*Identify*: to establish the distinguishing characteristic of; to select; to choose something from a number or group. The verb "identify" is also a level 1 verb; however, within the context of outcome 23 (lifelong learning), the verb "identify" conveys the ability to determine the additional knowledge, skills, and attitudes appropriate for professional practice, which is a level 4 task.

Other illustrative verbs at the *analysis* level include: break down, correlate, differentiate, discriminate, infer, and outline.

## Level 5—Synthesis

Synthesis refers to the ability to put together to form a new whole. This may involve the production of a unique communication, a plan of operation (research proposal), or a set of abstract relations (scheme for classifying information). Learning outcomes in this area stress creative behaviors and place major emphasis on the formulation of new patterns or structure.[e]

*Create*: to produce (as a work of art or of dramatic interpretation) along new or unconventional lines; to make or bring into existence something new.

*Design*: to conceive and plan out in the mind; to create, fashion, execute, or construct according to plan; to originate, draft, and work out, set up, or set forth.

*Specify*: to tell or state precisely or in detail. Although not usually considered a level 5 verb, when used with outcome 7 (experiments), the verb "specify" refers to the ability to determine which experiment or experiments are required. Drawing from a wide range of possibilities and then specifying the appropriate one(s) is a level 5 task.

*Explain*: to show the logical development or relationships of. "Explain" is also a level 2 verb when it simply means to make plain or understandable. Showing a logical development or relationships are level 5 tasks.

*Synthesize*: combine or put together by the composition or combination of parts or elements so as to form a whole; the combining of often varied and diverse ideas, forces, or factors into one coherent or consistent complex.

*Relate*: to show or establish a logical or causal connection between.

*Develop*: to open up; to cause to become more completely unfolded so as to reveal hidden or unexpected qualities or potentialities; to lay out (as a representation) into a clear, full, and explicit presentation. "Develop" is also a level 3 verb when—much like the verb "explain"—it means to make clear, plain, or understandable. For outcomes 10 (sustainability), 12 (risk and uncertainty), 17 (public policy), and 19 (globalization) develop requires synthesis.

*Plan*: to devise or project the realization or achievement of; to arrange the parts of.

*Compose*: to form by putting together two or more things, elements, or parts; to put together; to arrange in a fitting, proper, or orderly way.

*Integrate*: to make complete; to form into a more complete, harmonious, or coordinated entity, often by the addition or arrangement of parts or elements; to combine to form a more complete, harmonious, or coordinated entity; to incorporate (as an individual or group) into a larger unit or group.

*Construct*: to form, make, or create by combining parts or elements; to create by organizing ideas or concepts logically, coherently, or palpably; to draw with suitable instruments so as to fulfill certain specified conditions; to assemble separate and often disparate elements.

*Adapt*: to make suitable (for a new or different use or situation) by means of changes or modifications.

*Organize*: to arrange or constitute into a coherent unity in which each part has a special function or relation; to arrange by systematic planning and coordination of individual effort; to arrange elements into a whole of interdependent parts.

*Execute*: to put into effect; to carry out fully and completely.

Other illustrative verbs at the *synthesis* level include: anticipate, collaborate, combine, compile, devise, facilitate, generate, incorporate, modify, reconstruct, reorganize, revise, and structure.

## Level 6—Evaluation

Evaluation concerns the ability to judge the value of material for a given purpose. The judgments are to be based on definite criteria. These may be internal criteria (organization) or external criteria (relevance to the purpose) and the individual may determine the criteria or be given them. Learning outcomes in this area are highest in the cognitive hierarchy because they contain elements of all the other categories as well as conscious value judgments based on clearly defined criteria.[e]

*Evaluate*: to examine and judge concerning the worth, quality, significance, amount, degree, or condition of.

*Compare*: to examine the character or qualities of, especially for the purpose of discovering resemblances or differences. This definition is the same as for level 4; however, when used in context with the verb "evaluate" for outcome 8 (problem recognition and solving), the combined action requires evaluation and is a level 6 task.

*Appraise*: to judge and analyze the worth, significance or status of; especially to give a definitive expert judgment of the merit, rank, or importance of.

*Justify*: to prove or show to be just, desirable, warranted, or useful.

*Assess*: to analyze critically and judge definitively the nature, significance, status, or merit of; to determine the importance, size, or value of.

*Self-assess*: to personally or internally analyze critically and judge definitively the nature, significance, status, or merit of a personal trait. Outcome 23 (lifelong learning) uses the verb "self-assess" to convey the concept of introspective reflection.

Other illustrative verbs at the *evaluation* level include: compare and contrast, conclude, criticize, decide, defend, judge, and recommend.

## Cited Sources


a) Bloom, B. S., M. D. Englehart, E. J. Furst, W. H. Hill, and D. Krathwohl. 1956. *Taxonomy of Educational Objectives, the Classification of Educational Goals, Handbook I: Cognitive Domain.* David McKay, New York, NY.

b) Anderson, L., D. R. Krathwohl, P. W. Airasian, K. A. Cruikshank, R. E. Mayer, P. R. Pintrich, J. Raths, and M. C. Wittrock. 2001. *A Taxonomy of Learning, Teaching, and Assessment: A*

*Revision of Bloom's Taxonomy.* Addison Wesley Longman, Inc. New York, NY.

c) ASCE Levels of Achievement Subcommittee. 2005. *Levels of Achievement Applicable to the Body of Knowledge Required for Entry Into the Practice of Civil Engineering at the Professional Level,* Reston, VA, September. (http://www.asce.org/raisethebar)

d) *Webster's Third New International Dictionary, Unabridged.* Merriam-Webster, 2002. Available at http://unabridged.merriam-webster.com.

e) Gronlund, N. E. 1978. *Stating Objectives for Classroom Instruction,* 2nd Edition, Macmillan, New York, NY.

# APPENDIX G

# The Affective Domain of Bloom's Taxonomy

## Overview

The civil engineering body of knowledge (BOK) is central to the profession. The BOK is the necessary depth and breadth of knowledge, skills, and attitudes required of an individual entering the practice of civil engineering at the professional level in the 21st century. The levels of achievement are described in terms of a standard educational taxonomy, initiated by Bloom et al.[10]

Blooms taxonomy consists of three domains: cognitive, affective, and psychomotor. The cognitive domain refers to educational objectives that deal with the recall or recognition of knowledge and the development of intellectual abilities and skills. It is used exclusively herein to describe desirable civil engineering outcomes and levels of achievement.

The affective domain includes objectives that describe changes in interest, attitudes, and values and is an inseparable complement. Progress in the affective domain is described in terms of internalization of values. The affective domain provides a distinct and valuable vocabulary and set of concepts that are relevant to professional education.

> *Several outcomes already identified as important to the profession would be enhanced by descriptions in both the cognitive domain and the affective domain. Two examples are outcome 24 (professional and ethical responsibility) and outcome 16 (communication).*

These examples illustrate the value added by including an affective domain description in cases in which cognitive development alone does not cover the full scope of the outcome. The BOK2 Committee recommends that further work be undertaken in this area.

## Bloom's Taxonomy[a,b]

There are many developmental taxonomies. Each describes the same thing—the human person—and the educational process of human development. The purpose of a taxonomy is to break down this overall development process into smaller discernable "chunks" within which:

- Goals can be articulated
- Metrics of achievement can be constructed
- Achievement can be assessed.

Because any taxonomy attempts to describe the whole, constructing a hybrid of different taxonomies is ill-advised unless one is prepared to engage in educational research per se.

According to Bloom,[a] there are three domains:

- "...the cognitive domain ... includes those objectives [that] deal with the recall or recognition of knowledge and the development of intellectual abilities and skills."
- "...the affective domain ... includes objectives [that] describe changes in interest, attitudes, and values ..."
- the psychomotor domain, which includes "...the manipulative or motor-skill area."

The cognitive domain was found to be most amenable to easy study and formed the basis of the first Bloom-led study.[a] The second effort—by Krathwohl[b]—extended this into the affective domain without changing the cognitive domain. The third, or psychomotor domain, was in fact not recommended by Bloom[a] for further study, although it remains a distinct domain.[c]

In describing the affective domain, Krathwohl and others[b] adopted internalization as the basis of classification in this domain. This domain is easily summarized with the hybrid phrase "internalization of values and attitudes." Clearly, this is very different from the cognitive domain. The affective domain consists of five levels of increasing achievement, as illustrated in Table G-1.

Appendices prepared by Krathwohl and others[b] are very useful. Their appendices A and B summarize both the cognitive and affective domains. Descriptive phrases used in their Appendix A serve as examples to illustrate the affective domain and are quoted in Table G-2.

Perhaps the most compelling case for the relevance of the affective domain is the description of level 3, valuing: "This category will be found appropriate for many objectives that use the term 'attitude' (as well as, of course, 'value')."[b] Several more current sources and activities[d,e,f,g] provide additional discussion and example verbs for use in the affective domain originally developed by Krathwohl and others.[b]

As mentioned above there are many taxonomies, all seeking to describe the same thing: human development. The cognitive/affective divide is characteristic, but not universal. For example, the conceive-design-implement-operate (CDIO)[h] taxonomy is a more contemporary (2001) creation. It mingles these domains in a different manner, combining "professional skills and attitudes" and also "personal skills and attitudes" quite deep in the taxonomy. Because of this, combining parts from disparate taxonomies is not advised lest the fullness and unity of the object be lost.

## First Edition of the Body of Knowledge[i]

The first edition of the body of knowledge focuses on the knowledge, skills, and attitudes (KSA) required for the future civil engineer. There are 15 specific outcomes generally falling within the knowledge/skills arena. Beyond that, there is a significant discussion of attitudes. The following points are made:

- Attitudes are an essential component of the "what" dimension of the BOK.
- Attitudes are found to be integral parts of the BOK of other professions.

Table G-1. Levels and sublevels of achievement in the affective domain.[b]

| *Affective taxonomy* | | | |
|---|---|---|---|
| 1.0 | Receiving | 1.1 | Awareness |
| | | 1.2 | Willingness to receive |
| | | 1.3 | Controlled or selected attention |
| 2.0 | Responding | 2.1 | Acquiescence in responding |
| | | 2.2 | Willingness to respond |
| | | 2.3 | Satisfaction in response |
| 3.0 | Valuing | 3.1 | Acceptance of a value |
| | | 3.2 | Preference for a value |
| | | 3.3 | Commitment |
| 4.0 | Organization | 4.1 | Conceptualization of a value |
| | | 4.2 | Organization of a value system |
| 5.0 | Characterization by a value complex | 5.1 | Generalized set |
| | | 5.2 | Characterization |

- Studies point to the essential role of attitudes in individual and group achievement.
- Knowledge and skills are necessary, but not sufficient, for the fully professional engineer.
- Absent a proactive effort at the university level, many civil engineering students and young engineers are not likely to acquire such attitudes—or worse, are likely to acquire negative attitudes.

There are three levels of achievement in the first BOK report:[i] recognition, understanding, and ability. These are suggested to be used in describing achievement of outcomes 1–15. The report also suggested that attitudes be connected to the achievement of outcomes 1–15; however the mechanism is not clear.

## The Levels of Achievement Report[j]

In the levels of achievement (LOA) report, the three achievement levels—recognition, understanding, and ability—were deemed unworkable. The search for a replacement led this subcommittee to a survey of the assessment field and to the need for an established learning taxonomy. There are several. The subcommittee rejected the notion that ASCE could invent its own taxonomy. Following a review of extant taxonomies, Bloom's original taxonomy was found to be most useful. Specifically, the 15 outcomes were discussed in terms of Bloom's cognitive domain; and Bloom's six cognitive levels were recommended. The issue of attitudes and their connection to the 15 outcomes was not addressed, nor was the need for the affective domain, although the latter was noted.

By inference the LOA subcommittee found the cognitive domain of Bloom's Taxonomy sufficient for these original outcomes 1–15. The subcommittee's report was generally silent on addressing the need for or value of the affective domain.

Table G-2. Illustrative affective domain objectives excerpted from Krathwohl et al.,[b] Appendix A, "A Condensed Version of the Affective Domain."[a]

2.0 Responding

2.2 Willingness to respond

- Acquaints himself/herself with significant current issues in international, political, social, and economic affairs through voluntary reading and discussion.
- Acceptance of responsibility for his/her own health and for the protection of the health of others

2.3 Satisfaction in response

- Enjoyment of self-expression in music and in arts and crafts as another means of personal enrichment.
- Finds pleasure in reading for recreation.
- Takes pleasure in conversing with many different kinds of people

3.0 Valuing

3.1 Acceptance of a value

- Continuing desire to develop the ability to speak and write effectively.
- Grows in his/her sense of kinship with human beings of all nations.

3.2 Preference for a value

- Assumes responsibility for drawing reticent members of a group into conversation.
- Deliberately examines a variety of viewpoints on controversial issues with a view to forming opinions about them.
- Actively participates in arranging for the showing of contemporary artistic efforts.

3.3 Commitment

- Devotion to those ideas and ideals that are the foundations of democracy.
- Faith in the power of reason and in methods of experiment and discussion.

4.0 Organization

4.1 Conceptualization of a value

- Attempts to identify the characteristics of an art object that he/she admires.
- Forms judgments as to the responsibility of society for conserving human and material resources.

4.2 Organization of a value system

- Weighs alternative social policies and practices against the standards of the public welfare rather than the advantage of specialized and narrow interest groups.
- Develops a plan for regulating his/her rest in accordance with the demands of his/her activities.

5.0 Characterization by a value or value complex

5.1 Generalized set

- Readiness to revise judgments and to change behavior in the light of evidence.
- Judges problems and issues in terms of situations, issues, purposes, and consequences involved rather than in terms of fixed, dogmatic precepts or emotionally wishful thinking.

5.2 Characterization

- Develops for regulation of his/her personal and civic life a code of behavior based on ethical principles consistent with democratic ideals.
- Develops a consistent philosophy of life.

[a] This table excludes Section 1.0 and Section 2.1 as shown in Table G-1 because they are not appropriate for college education.

## The Curriculum Committee Report[k]

In parallel, the ASCE Committee on Academic Prerequisites for Professional Practice (CAP[3]) created the Curricula Committee. This group's work is quite comprehensive, examining the original BOK and the recommendations of the LOA effort. Specifically, the committee endorsed the original 15 outcomes and the use of Bloom's cognitive taxonomy. Regarding attitudes, this committee echoed the description in the first BOK report and supplemented it. It endorsed the importance of attitudes within the profession and echoed the idea of linking attitudes to the 15 outcomes.

Predictably, there were difficulties—attitudes that were not measurable (cognitive) outcomes; some attitudes demonstrably both "good" and "bad," depending on the context; no definitive list; and no metrics of assessment.

Among the conclusions of the Curriculum Committee:[k]

- Knowledge and skill are necessary, but not sufficient, for the practice of civil engineering.
- Professional attitudes can and should be learned.
- Attitudes cannot be taught, but can be "taught about."

The Curriculum Committee offered these recommendations:

- "Any use of this BOK to advance standardized measurements of attitude would be contrary to the committee's recommendations. When it comes to attitudes as part of the BOK, a flexible approach is in order.
- The committee believes that civil engineering departments and employers should adopt the approach that understanding the value and meaning of certain attitudes is an educational and developmental opportunity.
- The committee recommends that each employer and university civil and environmental engineering department select a set of constructive attitudes, possibly calling them professional attitudes. They may draw on the example list provided earlier or use other sources. They may choose to teach about the selected attitudes within the B+M/30&E process." (That is, during prelicensure formal education and experience.)

There is no recommendation here relative to what is clearly "affective" in Bloom's Taxonomy.

## Second Edition of the BOK

In the second edition of the BOK (this report), the 15 outcomes have been refined and expanded to 24. In so doing the cognitive domain has been used as the basis of the rubric, generally following the LOA suggestions. Significant progress has been made.

Regarding the BOK1 attitudes issue, the BOK2 Committee considered interpreting the KSA categories (knowledge, skills, and attitudes) as knowledge, skills, and abilities—consistent with an overall focus on the cognitive domain only. This has the advantage of being more readily measurable. But despite the appeal of using a different "A" word, it is not a synonym. After consideration, the BOK2 Committee rejected this substitution for two reasons:

- Abilities do not seem different from skills.
- The sense of "attitude" is entirely lost.

Interestingly, Bloom's cognitive domain denotes level 1 achievement as "knowledge;" while cognitive levels 2–6 are "intellectual abilities and skills."

Reflecting this, the BOK2 Committee has included an explicit, stand-alone attitude outcome: outcome 22 (attitudes). This outcome, like others, is described solely in terms of cognitive domain achievements. The cognitive domain provides a somewhat incomplete vocabulary for this outcome. There is overlap of the cognitive and affective domains, especially at the lower levels 1 and 2 of achievement. But at level 3, "valuing," the most obvious departure from the cognitive domain occurs. At level 3 and beyond, increasing affective achievement is uniquely described in terms of internalization of values and attitudes, a notion not relevant in the cognitive domain. Continued BOK2 Committee discussion indicates that there may be an affective dimension of achievement implicit in several identified outcomes.

## Conclusion: Two-Dimensional Outcomes

The BOK2 Committee concludes that there is value added in exploring an affective domain description of the present outcomes, to accompany the existing cognitive domain descriptions. There is nothing wrong with the cognitive domain; it is simply incomplete. The committee suggests a two-dimensional classification: cognitive and affective. This has the advantage of freeing some of the outcomes from a one-dimensional sense of achievement and allowing additional noncognitive verbs to enter the achievement descriptions as appropriate. A two-dimensional approach will add value to the description of the individual outcomes and add legitimacy in the sense of properly using the selected taxonomy.

The bottom line is this: the profession wants individuals who possess more than knowledge and skill.[1] The affective domain is one framework in which a more complete analysis and discussion can occur. Given the high and continuing interest in "affective" development, the BOK2 Committee recommends that this be explored, but not as part of the BOK2 Committee's efforts.

## Example Affective Domain Rubrics

Consider the two example rubrics, using the Affective Domain, that appear in Tables G-3 and G-4. These supplement and enrich the existing cognitive rubrics; they do not replace them. For illustrative purposes the cognitive rubric is replicated here without change. Several possibilities occur in the professional outcome category, but not all. There may be something useful in the foundational category such as a scientific respect for theory and observation; humanist values directed at needs; an internalization of the value inherent in diversity in teamwork; and the foundational basis of ethics. The committee has not attempted these, but feel that the point is best made in the examples selected.

The example rubrics are followed by commentaries on the affective domain portions.

Table G-3. Example rubric for a two-dimensional outcome—communication.

| *Outcome title* | *Level of affective achievement* | | | | |
|---|---|---|---|---|---|
| | *1 Receiving* | *2 Responding* | *3 Valuing* | *4 Organizing conceptualizing* | *5 Characterizing* |
| *To enter the practice of civil engineering at the professional level, an individual must be able to demonstrate this level of achievement* | | | | | |
| 16 Communication | ***Develop an awareness of*** the factors involved in effective verbal, written, virtual, and graphical communication. (B) | ***Discuss*** the factors involved in effective verbal, written, virtual, and graphical communications. (B) | ***Demonstrate*** a commitment to effective verbal, written, virtual, and graphical communications. (B) | ***Integrate*** principles from effective communications into work products. (E) | ***Discriminate*** between effective and ineffective communications. |

| *Outcome title* | *Level of cognitive achievement*[a] | | | | | |
|---|---|---|---|---|---|---|
| | *1 Knowledge* | *2 Comprehension* | *3 Application* | *4 Analysis* | *5 Synthesis* | *6 Evaluation* |
| *To enter the practice of civil engineering at the professional level, an individual must be able to demonstrate this level of achievement* | | | | | | |
| 16 Communication | ***List*** the characteristics of effective verbal, written, virtual, and graphical communications. (B) | ***Describe*** the characteristics of effective verbal, written, virtual, and graphical communications. (B) | ***Apply*** the rules of grammar and composition in verbal and written communications, properly cite sources, and ***use*** appropriate graphical standards in preparing engineering drawings. (B) | ***Organize*** and ***deliver*** effective verbal, written, virtual, and graphical communications. (B) | ***Plan, compose,*** and ***integrate*** the verbal, written, virtual, and graphical communication of a project to technical and nontechnical audiences. (E) | ***Evaluate*** the effectiveness of the integrated verbal, written, virtual, and graphical communication of a project to technical and nontechnical audiences. |

[a] Taken from Appendix I

Table G-4. Example rubric for a two-dimensional outcome—professional and ethical responsibility.

| Outcome title | *Level of affective achievement* | | | | |
|---|---|---|---|---|---|
| | *1*<br>*Receiving* | *2*<br>*Responding* | *3*<br>*Valuing* | *4*<br>*Organizing conceptualizing* | *5*<br>*Characterizing* |
| *To enter the practice of civil engineering at the professional level, an individual must be able to demonstrate this level of achievement* | | | | | |
| 24<br>Professional and ethical responsibility | ***Locate*** and ***identify*** the professional and ethical responsibilities of a civil engineer.<br>(B) | ***Discuss*** the professional and ethical responsibilities of a civil engineer.<br>(B) | ***Commit*** to the standards of professional and ethical responsibility for engineering practice.<br>(B) | ***Integrate*** professional and ethical standards for the engineer's own practice.<br>(B) | ***Display*** professional and ethical conduct in engineering practice.<br>(E) |

| Outcome title | *Level of cognitive achievement*[a] | | | | | |
|---|---|---|---|---|---|---|
| | *1*<br>*Knowledge* | *2*<br>*Comprehension* | *3*<br>*Application* | *4*<br>*Analysis* | *5*<br>*Synthesis* | *6*<br>*Evaluation* |
| *To enter the practice of civil engineering at the professional level, an individual must be able to demonstrate this level of achievement* | | | | | | |
| 24<br>Professional and ethical responsibility | ***List*** the professional and ethical responsibilities of a civil engineer.<br>(B) | ***Explain*** the professional and ethical responsibilities of a civil engineer.<br>(B) | ***Apply*** standards of professional and ethical responsibility to determine an appropriate course of action.<br>(B) | ***Analyze*** a situation involving multiple conflicting professional and ethical interests to determine an appropriate course of action.<br>(B) | ***Synthesize*** studies and experiences to foster professional and ethical conduct.<br>(E) | ***Justify*** a solution to an engineering problem based on professional and ethical standards and ***assess*** personal professional and ethical development.<br>(E) |

[a] Taken from Appendix I.

## Commentary on Affective Domain Portions of the Example Rubrics

### *Communications*

The cognitive domain approach to outcome 16 (communication) in Table G-3 clearly defines, demonstrates the importance of, and articulates the level of cognitive development of communications for civil engineers. The level of achievement at the time of licensure in the cognitive domain is synthesis (level 5 of 6) and requires the engineer to "*plan*, *compose*, and *integrate* verbal and graphical communications for both technical and nontechnical audiences." Upon completion of a baccalaureate education, the graduate is expected to "*organize* and *deliver* effective verbal, written, virtual, and graphical communications." To achieve this level, the engineer must complete the lower levels of achievement of knowledge, comprehension, and application. While this is necessary, without achievement in the affective domain, it is, in and of itself, insufficient.

The affective domain relates to the emotional component of learning and, in the case of communications, achievement is characterized by a degree of acceptance. A student and practitioner should acquire the intellectual skills needed to communicate effectively but unless the engineer internalizes those skills in a way that drives the engineer to want to communicate effectively, the educational and experience process falls short. The student and engineer must grow to see that effective communications are necessary for each and every communication and not only when being assessed or judged in some explicit manner.

Thus, the student must "*demonstrate* a commitment to effective verbal, written, virtual, and graphical communications." This affective domain level of achievement is valuing (affective level 3 of 5). To reach this level, the student must start with level 1 (receiving) by developing an awareness of the factors involved in effective communications and responding (affective level 2) by discussing the factors involved in effective communications. No doubt many students will successfully integrate principles of effective communications into work products (affective level 4)—into senior design reports, for example—but this level may not be achieved by all graduates until after graduation. Building upon the formal education, the engineer, through experience, continues to develop and, by the time of licensure, the engineer must be able to "*integrate* principles from effective communications into work products" (affective level 4).

Cognitive development and affective development are interrelated. An individual would not likely value and integrate effective communications if the individual had not achieved the intellectual skills (cognitive development) necessary to understand, organize, and compose effective communications.

### *Professional and Ethical Responsibility*

To enter the practice of civil engineering at the professional level, an engineer is expected to achieve the evaluation level (cognitive level 6) learning for outcome 24 (professional and ethical responsibility) in the cognitive domain as shown in Table G-4. At this highest level of cognitive learning, the civil engineer should be able to "*justify* a solution to an engineering problem based on professional and ethical standards and *assess* personal professional and ethical development." Despite this high level of learning, cognitive knowledge of ethical and professional responsibility seems lacking in the internalization and valuing of the knowledge. This is the crucial—and perhaps central—role of affective domain learning for this outcome.

Knowledge of professional and ethical responsibility should be internalized and valued in such a way that the civil engineering graduate does "*commit* to the standards of professional and ethical responsibility for engineering." This is an important early step in the development of a true professional working within appropriate ethical standards. Further, prior to entering the practice of civil engineering at the professional level, the engineer should have already demonstrated that he or she has "*integrated*" these standards into professional practice.

Level 5 of the affective domain for this outcome—characterizing—may be assessed positively for an engineer when he or she does "*display* professional and ethical standards in engineering practice" on a daily basis. This is the level of achievement expected for entry into civil engineering at the professional level.

Development within the affective domain is crucial to the effective practice of engineering. For example, an engineer who "*integrates* professional and ethical standards" into his or her practice (affective level 4) may not have knowledge (cognitive level 1) or comprehension (cognitive level 2) of a particular ethical or professional standard. However, because of his or her high affective level of achievement, that engineer would, through techniques of lifelong learning or the assistance of an expert, acquire the necessary level of achievement in the cognitive domain to make an appropriate decision.

## Recommendation for Future Work

An affective domain supplement to the BOK2 cognitive descriptions is possible and desirable. It is illustrated in this appendix by two examples and provides a richer description of BOK outcomes and achievement levels. Accordingly, the BOK2 Committee recommends that departments, schools, employers, and professionals develop these ideas more fully. The committee also recommends that ASCE continue this investigation more fully through $CAP^3$ activity beyond the present BOK2 development.

## Cited Sources


a) Bloom, B. S., M. D. Englehart, E. J. Furst., W. H. Hill, and D. R. Krathwohl. 1956. *The Taxonomy of Educational Objectives, The Classification of Educational Goals, Handbook I: Cognitive Domain.* David McKay Company, New York, NY.

b) Krathwohl, D. R., B. S. Bloom, and B. B. Masia. 1964. *The Taxonomy of Educational Objectives: The Classification of Educational Goals. Handbook II: Affective Domain*, David McKay Company, New York, NY.

c) Simpson, E. J. 1972. *The Classification of Educational Objectives in the Psychomotor Domain*, Gryphon House, Washington, DC.

d) http://www.flaguide.org/start/primerfull.php

e) http://www.acu.edu/academics/adamscenter/services/instructional/taxonomies.html#affective

f) http://classweb.gmu.edu/ndabbagh/Resources/Resources2/krathstax.htm

g) http://www.nwlink.com/~donclark/hrd/bloom.html

h) Crawley, E. F. 2001. *The CDIO Syllabus: A Statement of Goals for Undergraduate Engineering Education,* MIT CDIO Report #1, 2001. Available at http://www.cdio.org.

i) ASCE Body of Knowledge Committee. 2004. *Civil Engineering Body of Knowledge for the 21st Century,* Reston, VA, January. (http://www.asce.org/raisethebar)

j) ASCE Levels of Achievement Subcommittee. 2005. *Levels of Achievement Applicable to the Body of Knowledge Required for Entry Into the Practice of Civil Engineering at the Professional Level,* Reston, VA, September. (http://www.asce.org/raisethebar)

k) ASCE Curriculum Committee of the Committee on Academic Prerequisites for Professional Practice. 2006. *Development of Civil Engineering Curricula Supporting the Body of Knowledge for Professional Practice.* Reston, VA, December.

l) ASCE Task Committee to Plan a Summit on the Future of the Civil Engineering Profession. 2007. *The Vision for Civil Engineering in 2025,* Reston, VA, ASCE. (A PDF version is available, at no cost, from http://www.asce.org/Vision2025.pdf)

# APPENDIX H

# Relationship of ABET, BOK1, and BOK2 Outcomes

A review of the outcomes associated with the second edition of the ASCE body of knowledge may suggest that the number of outcomes has expanded greatly since the publication of the first edition,[3] and even more so relative to the outcomes listed in the ABET General Criteria for Baccalaureate Level Programs.

Tables H-1 and H-2 show that the number of outcomes has not expanded as much as casual observation might indicate. Relabeling and disaggregation of the BOK1 outcomes have increased the number of outcomes that are presented in this BOK2 report. However, the clarity and precision with which the outcomes are described have increased. This will assist the "raise the bar" intent of PS 465 and its role in the future education and prelicensure experience of civil engineers. The following paragraphs explain the process that led to the development of the 24 outcomes.

The outcomes approach resulting in BOK1 and BOK2 is the outcomes-based process implemented by EAC/ABET in its General Criteria for Baccalaureate Level Programs. The program criteria developed by the various engineering societies generally follows the format of the General Criteria for Baccalaureate Level Programs.

Section II.D.1 of the ABET *Accreditation Policy and Procedure Manual* provides the following useful definitions when using ABET accreditation criteria:

> *II.D.1.a. Program Educational Objective—Program educational objectives are broad statements that describe the career and professional accomplishments that the program is preparing graduates to achieve.*
>
> *II.D.1.b. Program Outcomes—Program outcomes are narrower statements that describe what students are expected to know and be able to do by the time of graduation. These relate to the skills, knowledge, and behaviors that students acquire in their matriculation through the program.*

The 11 outcomes expressed in Criterion 3(a) through Criterion 3(k) of the General Criteria for Baccalaureate Level Programs (hereinafter referred to as Criterion 3), are identified in the first column of Table H-1 and the third column of Table H-2. Table H-1 presents the genealogy of the outcomes from left to right—from the ABET criteria, through the BOK1 outcomes, to the BOK2 outcomes. Table H-2 is the reverse: the BOK2 outcomes are traced back to the ABET criteria.

The 11 Criterion 3 outcomes are universal for all engineering disciplines, and thus cannot be discipline-specific. In order to provide the specificity necessary to differentiate civil engineers from, for example, mechanical engineers or electrical engineers, ABET provides for discipline-specific program criteria. ASCE continually reviews and updates the civil engineering program criteria to specifically address the current educational needs of civil engineers. The civil engineering program criteria added further outcomes to the 11 from Criterion 3.

In addition to the 11 outcomes included in Criterion 3 and the program criteria, Criterion 5, "Curriculum" also stipulates some requirements for the engineering programs. The Criterion 5 requirements are not phrased as outcomes. Criterion 5(c) states that the professional component must include:

> *A general education component that complements the technical content of the curriculum and is consistent with the program and institution objectives.*

Concurrently with the initial development and launch of the Criterion 3 outcomes, ASCE was investigating the formal educational requirements for the professional practice of civil engineering in the future, which would become known as ASCE PS 465. ASCE was somewhat constrained by the EAC/ABET baccalaureate-level general criteria and the program criteria format that focused on the current conditions and, therefore, chose to broaden the scope to include a visionary aspect as envisioned by PS 465. This study by ASCE resulted in a series of reports by various subcommittees of what is now the ASCE Committee on the Academic Prerequisites for Professional Practice (CAP[3]). One of these reports was the original BOK report,[3] now known as the BOK1 report, published in January 2004.

The outcomes included in the BOK1 report subsumed the 11 Criterion 3 outcomes and added four more outcomes. These BOK1 outcomes are listed in the second column of both Tables H-1 and H-2. Table H-1 presents these outcomes in the numerical order in which they were presented in the BOK1 report. Table H-2 presents these outcomes as they are related to the outcomes of BOK2.

After the BOK1 was published and the civil engineering community digested and discussed the contents, three shortcomings became apparent. The first was the levels of achievement required for the individual outcomes. The three-level system used in the BOK1 report was ambiguous and imprecise. This was solved in the BOK2 report by utilizing the six levels inherent in the Bloom's Taxonomy of the cognitive domain, as discussed in Appendix F.

The second shortcoming noted was the broad scope of some of the outcomes, which was a reflection of the Criterion 3 format that had been followed. Outcomes had been grouped together to facilitate the ABET assessment process, but in reality, this proved to be a hindrance to flexibility and understanding and assessment of the outcomes. For example, as shown in the first row of Table H-1, mathematics, science, and engineering are grouped together as Criterion 3(a) in the EAC/ABET criteria and in BOK1. In BOK2, this outcome was disaggregated into four different outcomes. The primary reason for this disaggregation was that not all of the topics included in Criterion 3(a) require the same level of achievement in a typical civil engineering curriculum. By creating distinct outcomes, the BOK2 Committee concluded that the "mechanics" topic should have a higher level of achievement than "mathematics" in the typical curriculum. However, if within a particular civil engineering program the faculty decides that mathematics is more important than mechanics, they are free to raise the level of achievement required for mathematics, commensurate with their particular curriculum.

The third shortcoming that was expressed to the BOK2 Committee by the users of the BOK1 report was that some topics were missing or not adequately highlighted as a separate outcome. These topics are best seen in Table H-2. For example, outcome 2 (natural sciences) is not an EAC/ABET outcome, nor was it identified in BOK1 as an outcome. However, because BOK2 is looking toward the future, and the civil engineering profession is continually becoming more interdisciplinary, the committee believes that such other sciences as biology may become as important, or more important, than chemistry and physics to civil engineers of the future. The other outcome that fits into this category is outcome 17 (public policy).

The other subset of this third shortcoming is the BOK2 outcomes that were mentioned in BOK1 but were not elevated to the level of stand-alone outcomes. These include outcome 12 (risk and uncertainty), outcome 18 (business and public administration), and outcome 22 (attitudes). These outcomes—shown in the two tables—were deemed by the committee to be sufficiently important to justify their identification as stand-alone outcomes rather than as phrases in one of the other outcomes.

The final subset of this third shortcoming involves the new importance given to humanities and social sciences by inclusion in the BOK2 of outcome 3 (humanities) and outcome 4 (social sciences). In the BOK1 the general education component of the future civil engineer's undergraduate education was delegated to Criterion 5(c) of the baccalaureate-level general criteria. The consensus of the committee was that, in order for future civil engineers to realize their potential as technological leaders in a global community, the humanities and social sciences had to be elevated to the status of stand-alone outcomes.

As summarized in the tables, there are only two outcomes that are totally new relative to BOK1 and the EAC/ABET outcomes—namely, outcome 2 (natural sciences) and outcome 17 (public policy). There are an additional three outcomes that were included in a minor manner in the BOK1 but are not mentioned in a substantive manner in the EAC/ABET documents. These are outcome 12 (risk and uncertainty), outcome 18 (business and public administration), and outcome 22 (attitudes). In addition, "historical perspectives" has been included as a component of outcome 11 to complement contemporary issues, which appears as Criterion 3(j) and BOK1 outcome 10, both of which called for "a knowledge of contemporary issues."

## Table H-1. From the ABET program criteria outcomes to BOK2 outcomes.[a]

| *ABET Outcomes*[a] | *BOK1 Outcomes*[b] | *BOK2 Outcomes*[c] |
|---|---|---|
| (a) Mathematics, science, engineering | 1. Technical core | 1. Mathematics<br>2. Natural sciences<br>5. Materials science<br>6. Mechanics |
| (b) Experiments | 2. Experiments | 7. Experiments |
| (c) Design | 3. Design | 9. Design<br>10. Sustainability |
| | 3. Design | 12. Risk/uncertainty |
| (d) Multidisciplinary teams | 4. Multidisciplinary teams | 21. Teamwork |
| (e) Engineering problems | 5. Engineering problems | 8. Problem recognition and solving |
| (f) Professional and ethical responsibility | 6. Professional and ethical responsibility | 24. Professional and ethical responsibility |
| (g) Communication | 7. Communication | 16. Communication |
| (h) Impact of engineering | 8. Impact of engineering | 11. Contemporary issues and historical perspectives |
| (i) Lifelong learning | 9. Lifelong learning | 23. Lifelong learning |
| (j) Contemporary issues | 10. Contemporary issues | 11. Contemporary issues and historical perspectives<br>19. Globalization |
| (k) Engineering tools | 11. Engineering tools | 8. Problem recognition and solving |
| | 12. Specialized area related to civil engineering | 15. Technical specialization |
| Program Criteria for Civil and Similarly Named Engineering Programs | 13. Project management, construction, and asset management | 13. Project management |
| | 14. Business and public policy | 17. Public policy<br>18. Business and public administration |
| Program Criteria for Civil and Similarly Named Engineering Programs | 15. Leadership | 20. Leadership<br>22. Attitudes |
| EAC/ABET Criterion 5[d] | EAC/ABET Criterion 5[d] | 3. Humanities<br>4. Social sciences |
| Program Criteria for Civil and Similarly Named Engineering Programs | Program Criteria for Civil and Similarly Named Engineering Programs | 14. Breadth in civil engineering areas |

a) Short names[12]

b) Short names of outcomes appearing in the BOK1 report,[3] pp. 24–29

c) Short names from this report, Table 1, page 16

d) General education component

[a] General relationships are presented, not one-to-one mapping.

Table H-2. From the BOK2 outcomes to the ABET program criteria outcomes.[a]

| *BOK2 Outcomes[a]* | *BOK1 Outcomes[b]* | *ABET Outcomes[c]* |
|---|---|---|
| 1. Mathematics | 1. Technical core | (a) Mathematics, science, engineering |
| 2. Natural sciences | 1. Technical core | (a) Mathematics, science, engineering |
| 3. Humanities<br>4. Social sciences | EAC/ABET Criterion 5[d] | EAC/ABET Criterion 5[d] |
| 5. Materials science | 1. Technical core | (a) Mathematics, science, engineering |
| 6. Mechanics | 1. Technical core | (a) Mathematics, science, engineering |
| 7. Experiments | 2. Experiments | (b) Experiments |
| 8. Problem recognition and solving | 5. Engineering problems | (e) Engineering problems |
| 9. Design | 3. Design | (c) Design |
| 10. Sustainability | 3. Design | (c) Design |
| 11. Contemporary issues and historical perspectives | 8. Impact of engineering<br>10. Contemporary issues | (h) Impact of engineering<br>(j) Contemporary issues |
| 12. Risk/uncertainty | 3. Design | |
| 13. Project management | 13. Project management, construction, and asset management | Program Criteria for Civil and Similarly Named Engineering Programs |
| 14. Breadth in civil engineering areas | Program Criteria for Civil and Similarly Named Engineering Programs | Program Criteria for Civil and Similarly Named Engineering Programs |
| 15. Technical specialization | 12. Specialized area related to civil engineering | |
| 16. Communication | 7. Communication | (g) Communication |
| 17. Public policy | | |
| 18. Business and public administration | 14. Business and public policy | |
| 19. Globalization | 10. Contemporary issues | (j) Contemporary issues |
| 20. Leadership | 15. Leadership | Program Criteria for Civil and Similarly Named Engineering Programs |
| 21. Teamwork | 4. Multidisciplinary teams | (d) Multidisciplinary teams |
| 22. Attitudes | 15. Leadership | |
| 23. Lifelong learning | 9. Lifelong learning | (i) Lifelong learning |
| 24. Professional and ethical responsibility | 6. Professional and ethical responsibility | (f) Professional and ethical responsibility |

a) Short names from this report, Table 1, page 16

b) Short names of outcomes appearing in the BOK1 report,[3] Figure 5, pp. 24–29

c) Short names[12]

d) General education component

[a] General relationships are presented, not one-to-one mapping.

# APPENDIX I

# Body of Knowledge Outcome Rubric

Building on the recommendations of the Levels of Achievement Subcommittee,[9] the BOK2 Committee developed the outcome rubric.[14] The rubric communicates the following BOK characteristics:

- The 24 outcomes, categorized as foundational, technical, and professional and, within each category, organized in approximate pedagogical order
- The level of achievement that an individual must demonstrate for each outcome to enter the practice of civil engineering at the professional level
- For each outcome the portion to be fulfilled through the bachelor's degree, the portion to be fulfilled through the master's degree or equivalent (approximately 30 semester credits of acceptable graduate-level or upper-level undergraduate courses in a specialized technical area and/or professional practice area related to civil engineering), and the portion to be fulfilled through prelicensure experience

Key:

| | |
|---|---|
| B | Portion of the BOK fulfilled through the bachelor's degree |
| M/30 | Portion of the BOK fulfilled through the master's degree or the equivalent |
| E | Portion of the BOK fulfilled through prelicensure experience |
| | Achievement levels beyond minimums needed to enter professional practice |

| Outcome title | Level of cognitive achievement | | | | | |
|---|---|---|---|---|---|---|
| | *1 Knowledge* | *2 Comprehension* | *3 Application* | *4 Analysis* | *5 Synthesis* | *6 Evaluation* |
| | ▭ *To enter the practice of civil engineering at the professional level, an individual must be able to demonstrate this level of achievement* | | | | | |
| **Foundational Outcomes** | | | | | | |
| 1 Mathematics | ***Define*** key factual information related to mathematics through differential equations. (B) | ***Explain*** key concepts and problem-solving processes in mathematics through differential equations. (B) | ***Solve*** problems in mathematics through differential equations and ***apply*** this knowledge to the solution of engineering problems. (B) | ***Analyze*** a complex problem to determine the relevant mathematical principles and then apply that knowledge to solve the problem. | ***Create*** new knowledge in mathematics. | ***Evaluate*** the validity of newly created knowledge in mathematics. |
| 2 Natural sciences | ***Define*** key factual information related to calculus-based physics, chemistry, and one additional area of natural science. (B) | ***Explain*** key concepts and problem-solving processes in calculus-based physics, chemistry, and one additional area of natural science. (B) | ***Solve*** problems in calculus-based physics, chemistry, and one additional area of natural science and ***apply*** this knowledge to the solution of engineering problems. (B) | ***Analyze*** complex problems to determine the relevant physics, chemistry, and/or other areas of natural science principles and then apply that knowledge to solve the problem. | ***Create*** new knowledge in physics, chemistry, and/or others areas of natural science. | ***Evaluate*** the validity of newly created knowledge in physics, chemistry, and/or others areas of natural science. |
| 3 Humanities | ***Define*** key factual information from more than one area of the humanities. (B) | ***Explain*** key concepts from at least one area of the humanities and their relationship to civil engineering problems and solutions. (B) | ***Demonstrate*** the importance of the humanities in the professional practice of engineering (B) | ***Analyze*** a complex problem informed by issues raised in the humanities and apply these considerations in the development of a solution to the problem. | ***Create*** new knowledge in humanities. | ***Evaluate*** the validity of newly created knowledge in humanities. |

| Outcome title | Level of cognitive achievement | | | | | |
|---|---|---|---|---|---|---|
| | 1<br>*Knowledge* | 2<br>*Comprehension* | 3<br>*Application* | 4<br>*Analysis* | 5<br>*Synthesis* | 6<br>*Evaluation* |
| 4<br>Social sciences | ***Define*** key factual information from more than one area of social sciences.<br>(B) | ***Explain*** key concepts from at least one area of the social sciences and their relationship to civil engineering problems and solutions.<br>(B) | ***Demonstrate*** the incorporation of social sciences knowledge into the professional practice of engineering.<br>(B) | ***Analyze*** a complex problem incorporating social science knowledge and then apply that knowledge in the development of a solution to the problem. | ***Create*** new knowledge in social sciences. | ***Evaluate*** the validity of newly created knowledge in social sciences. |
| **Technical Outcomes** | | | | | | |
| 5<br>Materials science | ***Define*** key factual information related to materials science within the context of civil engineering.<br>(B) | ***Explain*** key concepts and problem-solving processes in materials science within the context of civil engineering.<br>(B) | ***Use*** knowledge of materials science to ***solve*** problems appropriate to civil engineering.<br>(B) | ***Analyze*** a complex problem to determine the relevant materials science principles, and then apply that knowledge to solve the problem. | ***Create*** new knowledge in materials science. | ***Evaluate*** the validity of newly created knowledge in materials science. |
| 6<br>Mechanics | ***Define*** key factual information related to solid and fluid mechanics.<br>(B) | ***Explain*** key concepts and problem-solving processes in solid and fluid mechanics.<br>(B) | ***Solve*** problems in solid and fluid mechanics.<br>(B) | ***Analyze*** and solve problems in solid and fluid mechanics.<br>(B) | ***Create*** new knowledge in mechanics. | ***Evaluate*** the validity of newly created knowledge in mechanics. |

| *Outcome title* | *Level of cognitive achievement* | | | | | |
|---|---|---|---|---|---|---|
| | *1 Knowledge* | *2 Comprehension* | *3 Application* | *4 Analysis* | *5 Synthesis* | *6 Evaluation* |
| 7 Experiments | ***Identify*** the procedures and equipment necessary to conduct civil engineering experiments in more than one of the technical areas of civil engineering. (B) | ***Explain*** the purpose, procedures, equipment, and practical applications of experiments spanning more than one of the technical areas of civil engineering. (B) | ***Conduct*** experiments in one or across more than one of the technical areas of civil engineering according to established procedures and ***report*** the results. (B) | ***Analyze*** the results of experiments and evaluate the accuracy of the results within the known boundaries of the tests and materials in or across more than one of the technical areas of civil engineering. (B) | ***Specify*** an experiment to meet a need, conduct the experiment, and analyze and *explain* the resulting data. (M/30) | ***Evaluate*** the effectiveness of a designed experiment in meeting an ill-defined real-world need. |
| 8 Problem recognition and solving | ***Identify*** key factual information related to engineering problem recognition, problem solving, and applicable engineering techniques and tools. (B) | ***Explain*** key concepts related to problem recognition, problem articulation, and problem-solving processes, and how engineering techniques and tools are applied to solve problems. (B) | ***Develop*** problem statements and ***solve*** well-defined fundamental civil engineering problems by ***applying*** appropriate techniques and tools. (B) | ***Formulate*** and solve an ill-defined engineering problem appropriate to civil engineering by ***selecting*** and applying appropriate techniques and tools. (M/30) | ***Synthesize*** the solution to an ill-defined engineering problem into a broader context that may include public policy, social impact, or business objectives. | ***Compare*** the initial and final problem statements, the effectiveness of alternative techniques and tools, and ***evaluate*** the effectiveness of the solution. |

| Outcome title | Level of cognitive achievement | | | | | |
|---|---|---|---|---|---|---|
| | 1<br>*Knowledge* | 2<br>*Comprehension* | 3<br>*Application* | 4<br>*Analysis* | 5<br>*Synthesis* | 6<br>*Evaluation* |
| 9<br>Design | ***Define*** engineering design; ***list*** the major steps in the engineering design process; and ***list*** constraints that affect the process and products of engineering design.<br>(B) | ***Describe*** the engineering design process; ***explain*** how real-world constraints affect the process and products of engineering design.<br>(B) | ***Apply*** the design process to meet a well-defined set of requirements and constraints.<br>(B) | ***Analyze*** a system or process to determine requirements and constraints.<br>(B) | ***Design*** a system or process to meet desired needs within such realistic constraints as economic, environmental, social, political, ethical, health and safety, constructability, and sustainability.<br>(B) | ***Evaluate*** the design of a complex system, component, or process and ***assess*** compliance with customary standards of practice, user's and project's needs, and relevant constraints.<br>(E) |
| 10<br>Sustainability | ***Define*** key aspects of sustainability relative to engineering phenomena, society at large, and its dependence on natural resources; and relative to the ethical obligation of the professional engineer.<br>(B) | ***Explain*** key properties of sustainability, and their scientific bases, as they pertain to engineered works and services.<br>(B) | ***Apply*** the principles of sustainability to the design of traditional and emergent engineering systems.<br>(B) | ***Analyze*** systems of engineered works, whether traditional or emergent, for sustainable performance.<br>(E) | ***Design*** a complex system, process, or project to perform sustainably. ***Develop*** new, more sustainable technology. ***Create*** new knowledge or forms of analysis in areas in which scientific knowledge limits sustainable design. | ***Evaluate*** the sustainability of complex systems, whether proposed or existing. |

| *Outcome title* | *Level of cognitive achievement* | | | | | |
|---|---|---|---|---|---|---|
| | *1 Knowledge* | *2 Comprehension* | *3 Application* | *4 Analysis* | *5 Synthesis* | *6 Evaluation* |
| 11 Contemporary issues and historical perspectives | ***Identify*** economic, environmental, political, societal, and historical aspects in engineering. (B) | ***Describe*** the influence of historical and contemporary issues on the identification, formulation, and solution of engineering problems and ***describe*** the influence of engineering solutions on the economy, environment, political landscape, and society. (B) | Drawing upon a broad education, ***explain*** the impact of historical and contemporary issues on the identification, formulation, and solution of engineering problems and ***explain*** the impact of engineering solutions on the economy, environment, political landscape, and society. (B) | ***Analyze*** the impact of historical and contemporary issues on the identification, formulation, and solution of engineering problems and ***analyze*** the impact of engineering solutions on the economy, environment, political landscape, and society. (E) | ***Synthesize*** the impacts and relationships among engineering and economic, environmental, political, societal, and historical issues. | ***Evaluate*** the impacts and relationships among engineering and historical, contemporary, and emerging issues. |
| 12 Risk and uncertainty | ***Recognize*** uncertainties in data and knowledge and ***list*** those relevant to engineering design. (B) | ***Distinguish*** between uncertainties that are data-based and those that are knowledge-based and ***explain*** the significance of those uncertainties on the performance and safety of an engineering system. (B) | ***Apply*** the principles of probability and statistics to ***solve*** problems containing uncertainties. (B) | ***Analyze*** the loading and capacity, and the effects of their respective uncertainties, for a well-defined design and ***illustrate*** the underlying probability of failure (or nonperformance) for a specified failure mode. (E) | ***Develop*** criteria (such as required safety factors) for the ill-defined design of an engineered system within an acceptable risk measure. | ***Appraise*** a multicomponent system and ***evaluate*** its quantitative risk measure, taking into account the occurrence probability of an adverse event and its potential consequences caused by failure. |

| Outcome title | Level of cognitive achievement | | | | | |
|---|---|---|---|---|---|---|
| | 1 *Knowledge* | 2 *Comprehension* | 3 *Application* | 4 *Analysis* | 5 *Synthesis* | 6 *Evaluation* |
| 13 Project management | ***List*** key management principles. (B) | ***Explain*** what a project is and the key aspects of project management. (B) | ***Develop*** solutions to well-defined project management problems. (B) | ***Formulate*** documents to be incorporated into the project plan. (E) | ***Create*** project plans. | ***Evaluate*** the effectiveness of a project plan. |
| 14 Breadth in civil engineering areas | ***Define*** key factual information related to at least four technical areas appropriate to civil engineering. (B) | ***Explain*** key concepts and problem-solving processes in at least four technical areas appropriate to civil engineering. (B) | ***Solve*** problems in or across at least four technical areas appropriate to civil engineering. (B) | ***Analyze*** and solve well-defined engineering problems in at least four technical areas appropriate to civil engineering. (B) | ***Create*** new knowledge that spans more than one technical area appropriate to civil engineering. | ***Evaluate*** the validity of newly created knowledge that spans more than one technical area appropriate to civil engineering. |
| 15 Technical specialization | ***Define*** key aspects of advanced technical specialization appropriate to civil engineering. (B) | ***Explain*** key concepts and problem-solving processes in a traditional or emerging specialized technical area appropriate to civil engineering. (M/30) | ***Apply*** specialized tools, technology, or technologies to ***solve*** simple problems in a traditional or emerging specialized technical area of civil engineering. (M/30) | ***Analyze*** a complex system or process in a traditional or emerging specialized technical area appropriate to civil engineering. (M/30) | ***Design*** a complex system or process or ***create*** new knowledge or technologies in a traditional or emerging advanced specialized technical area appropriate to civil engineering. (M/30) | ***Evaluate*** the design of a complex system or process, or ***evaluate*** the validity of newly created knowledge or technologies in a traditional or emerging advanced specialized technical area appropriate to civil engineering. (E) |

| Outcome title | Level of cognitive achievement | | | | | |
|---|---|---|---|---|---|---|
| | 1 Knowledge | 2 Comprehension | 3 Application | 4 Analysis | 5 Synthesis | 6 Evaluation |
| **Professional Outcomes** | | | | | | |
| 16 Communication | ***List*** the characteristics of effective verbal, written, virtual, and graphical communications. (B) | ***Describe*** the characteristics of effective verbal, written, virtual, and graphical communications. (B) | ***Apply*** the rules of grammar and composition in verbal and written communications, properly cite sources, and ***use*** appropriate graphical standards in preparing engineering drawings. (B) | ***Organize*** and ***deliver*** effective verbal, written, virtual, and graphical communications. (B) | ***Plan, compose,*** and ***integrate*** the verbal, written, virtual, and graphical communication of a project to technical and nontechnical audiences. (E) | ***Evaluate*** the effectiveness of the integrated verbal, written, virtual, and graphical communication of a project to technical and nontechnical audiences. |
| 17 Public policy | ***Describe*** key factual information related to public policy. (B) | ***Discuss*** and ***explain*** key concepts and processes involved in public policy. (B) | ***Apply*** public policy process techniques to simple public policy problems related to civil engineering works. (E) | ***Analyze*** real-world public policy problems on civil engineering projects. | ***Develop*** public policy recommendations, and ***create*** or ***adapt*** a system to a real-world situation on civil engineering work programs. | ***Evaluate*** the effectiveness of a public policy in a complex, real-world situation associated with large-scale civil engineering initiatives. |
| 18 Business and public administration | ***List*** key factual information related to business and public administration. (B) | ***Explain*** key concepts and processes used in business and public administration. (B) | ***Apply*** business and public administration concepts and processes. (E) | ***Analyze*** real-world problems involving business or public administration. | ***Create*** or ***adapt*** a system of business or public administration to meet a real-world need. | ***Evaluate*** a system of business or public administration in a complex, real-world situation. |

| *Outcome title* | *Level of cognitive achievement* | | | | | |
|---|---|---|---|---|---|---|
| | *1 Knowledge* | *2 Comprehension* | *3 Application* | *4 Analysis* | *5 Synthesis* | *6 Evaluation* |
| 19 Globalization | ***Describe*** globalization processes and their impact on professional practice across cultures, languages, or countries. (B) | ***Explain*** global issues related to professional practice, infrastructure, environment, and service populations (as they arise across cultures, languages, or countries). (B) | ***Organize, formulate,*** and ***solve*** engineering problems within a global context. (B) | ***Analyze*** engineering works and services in order to function at a basic level in a global context. (E) | ***Develop*** criteria and guidelines to address global issues. | ***Evaluate*** different criteria and guidelines in addressing global issues. |
| 20 Leadership | ***Define*** leadership and the role of a leader; ***list*** leadership principles and attitudes. (B) | ***Explain*** the role of a leader and leadership principles and attitudes. (B) | ***Apply*** leadership principles to direct the efforts of a small, homogenous group. (B) | ***Organize*** and ***direct*** the efforts of a group. (E) | ***Create*** a new organization to accomplish a complex task. | ***Evaluate*** the leadership of an organization. |
| 21 Teamwork | ***Define*** and ***list*** the key characteristics of effective intradisciplinary and multidisciplinary teams. (B) | ***Explain*** the factors affecting the ability of intradisciplinary and multidisciplinary teams to function effectively. (B) | ***Function*** effectively as a member of an intradisciplinary team. (B) | ***Function*** effectively as a member of a multidisciplinary team. (E) | ***Organize*** an intradisciplinary or multidisciplinary team. | ***Evaluate*** the composition, organization, and performance of an intradisciplinary or multidisciplinary team. |

| Outcome title | Level of cognitive achievement | | | | | |
|---|---|---|---|---|---|---|
| | 1 *Knowledge* | 2 *Comprehension* | 3 *Application* | 4 *Analysis* | 5 *Synthesis* | 6 *Evaluation* |
| 22 Attitudes | ***List*** attitudes supportive of the professional practice of civil engineering. (B) | ***Explain*** attitudes supportive of the professional practice of civil engineering. (B) | ***Demonstrate*** attitudes supportive of the professional practice of civil engineering. (E) | ***Analyze*** a complex task to determine which attitudes are most conducive to its effective accomplishment. | ***Create*** an organizational structure that maintains/fosters the development of attitudes conducive to task accomplishment. | ***Evaluate*** the attitudes of key members of an organization and ***assess*** the effect of their attitudes on task accomplishment. |
| 23 Lifelong learning | ***Define*** lifelong learning. (B) | ***Explain*** the need for lifelong learning and ***describe*** the skills required of a lifelong learner. (B) | ***Demonstrate*** the ability for self-directed learning. (B) | ***Identify*** additional knowledge, skills, and attitudes appropriate for professional practice. (E) | ***Plan*** and ***execute*** the acquisition of required expertise appropriate for professional practice. (E) | ***Self-assess*** learning processes and ***evaluate*** those processes in light of competing and complex real-world alternatives. |
| 24 Professional and ethical responsibility | ***List*** the professional and ethical responsibilities of a civil engineer. (B) | ***Explain*** the professional and ethical responsibilities of a civil engineer. (B) | ***Apply*** standards of professional and ethical responsibility to determine an appropriate course of action. (B) | ***Analyze*** a situation involving multiple conflicting professional and ethical interests to determine an appropriate course of action. (B) | ***Synthesize*** studies and experiences to foster professional and ethical conduct. (E) | ***Justify*** a solution to an engineering problem based on professional and ethical standards and ***assess*** personal professional and ethical development. (E) |

# APPENDIX J

# Explanations of Outcomes

## Introduction

The BOK2 Committee created the following explanations for each of the 24 outcomes. These explanations are designed to help faculty who teach aspiring civil engineers and practitioners who supervise, coach, mentor, educate, train, and inspire prelicensure civil engineers. The explanations will also aid civil engineering students and engineer interns who are preparing for entry into the professional practice of civil engineering. To reiterate, explanations are to be helpful—they are not prescriptive. Outcomes paired with explanations provide what the committee views as a desirable deliverable for stakeholders; Bloom's Taxonomy-based outcomes relying on active verbs, with each outcome supported by a descriptive and illustrative explanation.

Outcomes are viewed as applicable over a long period—years, for example. In contrast, some illustrative topics mentioned in the explanations will be ephemeral, requiring modification in response to stakeholder needs, technological advances, and other changes.

The format used for the explanations enables the reader to readily move from one outcome to another because formats are identical. The format for each explanation begins with an overview that presents the rationale for the outcome and defines terms, as needed.

The overview is followed by a section—denoted by "B"—that states the minimum level of achievement to be fulfilled through the bachelor's degree. The level of achievement is taken directly from the rubric. An L1, L2, L3, L4, L5, or L6 is included to reiterate, respectively, the following Bloom's Taxonomy level of achievement that is to be accomplished: knowledge, comprehension, application, analysis, synthesis, and evaluation. The "B" section goes on to offer ideas on curricular and, in some cases, co-curricular and extracurricular ways to enable the aspiring civil engineer to reach the required levels of achievement.

As appropriate for the outcome, the "B" section is followed by an "M/30" (master's degree or equivalent) and/or an "E" (experience) section. As with the "B" section, these sections offer ideas on how an individual, within his or her courses or during his or her prelicensure experience, can attain the necessary minimum levels of achievement.

# FOUNDATIONAL OUTCOMES

# Outcome 1: Mathematics

## Overview

Mathematics deals with the science of structure, order, and relation that has evolved from counting, measuring, and describing the shapes of objects. It uses logical reasoning and quantitative calculation. Since the 17th century mathematics has been an indispensable adjunct to the physical sciences and technology and is considered the underlying language of science. The principal branches of mathematics relevant to civil engineering are algebra, analysis, arithmetic, geometry, calculus, numerical analysis, optimization, probability, set theory, statistics, and trigonometry.

All areas of civil engineering rely on mathematics for the performance of quantitative analysis of engineering systems. A technical core of knowledge and breadth of coverage in mathematics, and the ability to apply it to solve engineering problems, are essential skills for civil engineers.

B: ***Solve*** **problems in mathematics through differential equations and *apply* this knowledge to the solution of engineering problems.** (L3) The mathematics required for civil engineering practice must be learned at the undergraduate level and should prepare students for subsequent courses in engineering curricula.

# FOUNDATIONAL OUTCOMES

# Outcome 2: Natural Sciences

## Overview

Underlying the professional role of the civil engineer as the master integrator and technical leader is a firm foundation in the natural sciences. Physics and chemistry are two disciplines of the natural sciences that have historically served as basic foundations. Additional disciplines of natural science are also assuming stronger roles within civil engineering.

Physics is concerned with understanding the structure of the natural world and explaining natural phenomena in a fundamental way in terms of elementary principles and laws. The fundamentals of physics are mechanics and field theory. Mechanics is concerned with the equilibrium and motion of particles or bodies under the action of given forces. The physics of fields encompasses the origin, nature, and properties of gravitational, electromagnetic, nuclear, and other force fields. Taken together, mechanics and field theory constitute the most fundamental approach to an understanding of natural phenomena that science offers. Physics is characterized by accurate instrumentation, precision of measurement, and the expression of its results in mathematical terms. Many areas of civil engineering rely on physics for understanding the underlying governing principles and for obtaining solutions to problems. A technical core of knowledge and breadth of coverage in physics, and the ability to apply it to solve engineering problems, are essential for civil engineers.

Chemistry is the science that deals with the properties, composition, and structure of substances (elements and compounds), the reactions and transformations they undergo, and the energy released or absorbed during those processes. Chemistry is concerned with atoms as building blocks, everything in the material world, and all living things. Branches of chemistry include inorganic, organic, physical, and analytical chemistry; biochemistry; electrochemistry; and geochemistry. Some areas of civil engineering—especially environmental engineering and construction materials—rely on chemistry for explaining phenomena and obtaining solutions to problems. A technical core of knowledge and breadth of coverage in chemistry is necessary for individuals to solve related problems in civil and environmental engineering.

Additional breadth in such natural science disciplines as biology, ecology, geology/geomorphology, et cetera is required to prepare the civil engineer of the future. Increased exposure to or emphasis on biological systems, ecology, sustainability, and nanotechnology is expected to occur in the 21st century. Civil engineers should have the basic scientific literacy that will enable them to be conversant with technical issues pertaining to environmental systems, public health and safety, durability of construction materials, and other such subjects. A technical core of knowledge and breadth of coverage in an area of science other than mathematics, physics, and chemistry is required to prepare future civil engineers.

**B: *Solve* problems in calculus-based physics, chemistry, and one additional area of natural science and *apply* this knowledge to the solution of engineering problems.** (L3) The physics, chemistry, and breadth in natural sciences required for civil engineering practice must be learned at the undergraduate level and should prepare students for subsequent courses in engineering and engineering practice.

# FOUNDATIONAL OUTCOMES

# Outcome 3: Humanities

## Overview[1]

To be effective, professional civil engineers must be critical thinkers and possess the ability to raise vital questions and problems and then formulate them clearly and appropriately. They must gather and assess relevant information, use abstract ideas to interpret the information effectively, and come to well-reasoned conclusions and solutions, testing them against relevant criteria and standards. Professional civil engineers must think openmindedly within alternative systems of thought, recognizing and assessing, as need be, the assumptions, implications, and practical consequences of their work. They must be informed not only by mathematics and the natural and social sciences, but by the humanities—the disciplines that study the human aspects of the world, including philosophy, history, literature, the visual and performing arts, language, and religion. Humanities are academic disciplines that use critical or speculative methods to study the human condition. This outcome is intended to guide students to understand the importance of the humanities on the professional practice of engineering. This understanding is critical to the professional delivery of service to people.

The formal education process sets the stage for professional achievement. Engineering practice often includes aesthetic, ethical, and historical considerations and other elements of the humanities. Therefore, engineers must be able to recognize and incorporate such human elements into the development and evaluation of solutions to engineering and societal problems. Continued development of professional competence must come from lifelong learning, mentorship from senior engineers, practical experience, and involvement in the local community grounded on a firm foundation in, and recognition of the importance of, the humanities.

**B: *Demonstrate* the importance of the humanities in the professional practice of engineering.** (L3) The formal education process at the undergraduate level must include the humanities in order for the student to develop an appreciation of their importance in developing engineering solutions. All students cannot study all of the humanities; rather, students first must be able to recognize and identify factual information from more than one area of the humanities. Students should be able to explain concepts in at least one area of humanities in order for them to explain how this can inform and impact their engineering decisions. Students should be

1. See Appendix K for additional ideas and information about the humanities.

able to apply their knowledge of the humanities by demonstrating the importance of the humanities on the professional practice of engineering. Examples of opportunities to demonstrate this ability include incorporating application of philosophy into engineering ethics, the visual arts into the aesthetics of structures, language into the globalization of engineering, and history in the study of the past accomplishments of society through civil engineering.

# FOUNDATIONAL OUTCOMES

# Outcome 4: Social Sciences

## Overview[1]

Engineering services are delivered to society through social mechanisms and institutions. The social sciences are the systematic study of these social phenomena. Example disciplines include economics, political science, sociology, and psychology. (Note that some studies in history are categorized as social sciences.) Social sciences are scientific, quantitative, analytical, and data-driven and use the scientific method, including both qualitative and quantitative methods. Professional civil engineers must work within a social framework; understanding it is foundational to effective professionalism, alongside the three other foundational areas—mathematics, the natural sciences, and the humanities. This outcome is intended to guide students to make connections between their technical education and their education in the social sciences. Effective delivery of professional service depends critically upon these connections.

The formal education process sets the stage for individuals to become effective professionals. In practice, virtually all projects and design work involve varying degrees of integration of social sciences knowledge, including the economic and sociopolitical aspects. Engineers must be able to recognize and incorporate these considerations into the development, delivery, and evaluation of solutions to engineering problems. Continued development of professional competence must come from lifelong learning, mentorship from senior engineers, and practical experience, grounded on a firm foundation in, and recognition of, the importance of the social sciences and advances in them.

**B: *Demonstrate* the incorporation of social sciences knowledge into the professional practice of engineering.** (L3) The formal education process at the undergraduate level must include an introduction to social sciences in order for the student to develop an appreciation of their importance in the development of engineering solutions. All students cannot master all of the social sciences; rather, students first must be able to recognize and identify factual information in more than one area of social science. Students should be able to explain the concepts in at least one area of social science in order to explain how this area of social science can inform their engineering decisions. Students should be able to apply their knowledge of social sciences by demonstrating its incorporation into the professional practice of engineering. Examples of knowledge from social sciences that might be applied in engineering include economic, safety and

1. See Appendix K for additional ideas and information about the social sciences.

security, or environmental considerations. Examples of opportunities to demonstrate this ability include incorporating the application of social sciences in such engineering courses as transportation, environmental engineering, capstone, or major design experience.

# TECHNICAL OUTCOMES

# Outcome 5: Materials Science

## Overview

Civil engineering includes elements of materials science. Construction materials with broad applications in civil engineering include such ceramics as Portland cement concrete and hot mix asphalt concrete, such metals as steel and aluminum, and polymers and fibers. An understanding of materials science also is required for the treatment of hazardous wastes utilizing membranes and filtration. Infrastructure often requires repair, rehabilitation, or replacement due to degradation of materials. The civil engineer is responsible for specifying appropriate materials. The civil engineer should have knowledge of how materials systems interact with the environment so that durable materials that can withstand aggressive environments can be specified as needed. This includes the understanding of materials at the macroscopic and microscopic levels.

A technical core of knowledge and breadth of coverage in materials science appropriate to civil engineering is necessary for individuals to solve a variety of civil engineering problems.

**B: *Use* knowledge of materials science to *solve* problems appropriate to civil engineering.** (L3) The materials science required for civil engineering practice must be learned at the undergraduate level and should prepare students for subsequent courses in engineering curricula.

# TECHNICAL OUTCOMES

# Outcome 6: Mechanics

## Overview

In its original sense, mechanics refers to the study of the behavior of systems under the action of forces. Mechanics is subdivided according to the types of systems and phenomena involved. An important distinction is based on the size of the system. The Newtonian laws of classical mechanics can adequately describe those systems that are large enough, including those encountered in most civil engineering areas. On the other hand, the concepts and mathematical methods of quantum mechanics must be employed to describe the behavior of such microscopic systems as molecules, atoms, and nuclei. Mechanics may also be classified as nonrelativistic or relativistic, the latter applying to systems with material velocities comparable to the velocity of light. This distinction pertains to both classical and quantum mechanics. Finally, statistical mechanics uses the methods of statistics for both classical and quantum systems containing very large numbers of similar subsystems to obtain their large-scale properties.

Mechanics in civil engineering encompasses the mechanics of continuous and particulate solids subjected to load, and the mechanics of fluid flow through pipes, channels, and porous media. Areas of civil engineering that rely heavily on mechanics are structural engineering, geotechnical engineering, pavement engineering, and water resource systems.

A technical core of knowledge and breadth of coverage in solid and fluid mechanics, and the ability to apply it to solve engineering problems, are essential for civil engineers.

**B: *Analyze* and solve problems in solid and fluid mechanics.** (L4) The mechanics required for civil engineering practice must be learned at the undergraduate level and should prepare students for subsequent courses in engineering curricula.

# TECHNICAL OUTCOMES

# Outcome 7: Experiments

## Overview

Experiment can be defined as "an operation or procedure carried out under controlled conditions in order to discover an unknown effect or law, to test or establish a hypothesis, or to illustrate a known law." [46]

Civil engineers frequently design and conduct field and laboratory studies, gather data, create numerical simulations and other models, and then analyze and interpret the results. The licensed civil engineer should be able to develop and conduct experiments and analyze results of experiments that may incorporate or span more than one current and/or emerging technical area appropriate to civil engineering. Inquiry-based learning emphasizing the method of discovery develops critical thinking skills necessary in learning the experimental process. Critical thinking also helps develop engineering judgment, necessary in interpreting and analyzing results of experiments.

**B: *Analyze* the results of experiments and evaluate the accuracy of the results within the known boundaries of the tests and materials in or across more than one of the technical areas of civil engineering.** (L4) Individuals should be familiar with the purpose, procedures, equipment, and practical applications of experiments spanning more than one of the technical areas of civil engineering. They should be able to conduct experiments, report results, and analyze results in accordance with the applicable standards in or across more than one technical area. In this context, experiments may include field and laboratory studies, virtual experiments, and numerical simulations.

**M/30: *Specify* an experiment to meet a need, conduct the experiment, and analyze and *explain* the resulting data.** (L5) The post baccalaureate experience related to experiments should prepare individuals to formulate, conduct, and analyze experiments on the basis of a specific need. This requires in-depth familiarity with the need as well as with the available and possible experimental tools. Individuals at this level are also expected to be familiar with the limitations of the experimental methods with which they deal. As at the baccalaureate level, experiments in this context may include field and laboratory studies, virtual experiments, and numerical simulations.

# TECHNICAL OUTCOMES

# Outcome 8: Problem Recognition and Solving

## Overview

Civil engineering problem solving consists of identifying engineering problems, obtaining background knowledge, understanding existing requirements and/or constraints, articulating the problem through technical communication, formulating alternative solutions—both routine and creative—and recommending feasible solutions.

The approach to problem solving should use a combination of criteria employing critical thinking and the desire to discover. The knowledge and abilities included in this outcome should not be limited to those necessary to identify and solve existing problems, but extended to include those required to anticipate opportunities in which knowledge and abilities can be applied for the common good. Problem recognition and problem solving are learning processes in which various tools enhance these learning processes and aid in discovering appropriate solutions.

Appropriate techniques and tools—including information technology, contemporary analysis and design methods, and design codes and standards to complement knowledge of fundamental concepts—are required to solve engineering problems. Problem solving also involves the ability to select the appropriate tools as a method to promote or increase the future learning ability of individuals.

**B: *Develop* problem statements and *solve* well-defined fundamental civil engineering problems by *applying* appropriate techniques and tools.** (L3) Civil engineers should be familiar with factual information related to engineering problem recognition and problem-solving processes. Additionally, civil engineers should be able to explain key concepts related to engineering problem recognition, articulation, and solving. Engineering problem solving tools should be taught with appropriate fundamental engineering technologies. These tools range from the simple to the complex and an understanding of the quality of the data obtained from or required for the use of the engineering tools is necessary to define the capability and/or value of each tool.

**M/30: *Formulate* and solve an ill-defined engineering problem appropriate to civil engineering by *selecting* and applying appropriate techniques and tools.** (L4) Prior to licensing, civil engineers must be able to analyze and solve engineering problems with poorly defined or incomplete parameters. Advanced level problem-solving knowledge and abilities acquired through post baccalaureate education are required. At this level, civil engineers are expected to anticipate and identify problems and opportunities in various systems and environments.

The engineering tools studied at this level and above will most likely be in specialized areas of professional practice. These tools usually require a sufficient fundamental knowledge of various technologies and technical breadth in order to select and organize their use for problem solving or integration into design problems. Civil engineers must also comprehend the limitations of the selected tools and/or computer model simulations.

# TECHNICAL OUTCOMES

# Outcome 9: Design

## Overview

Design is an iterative process that is often creative and involves discovery and the acquisition of knowledge. Such activities as problem definition, the selection or development of design options, analysis, detailed design, performance prediction, implementation, observation, and testing are parts of the engineering design process.

Design problems are often ill defined. Thus defining the scope and design objectives and identifying the constraints governing a particular problem are essential to the design process. The design process is open-ended and involves a number of likely correct solutions, including innovative approaches. Thus successful design requires critical thinking, an appreciation of the uncertainties involved, and the use of engineering judgment. Such considerations as risk assessment, societal and environmental impact, standards, codes, regulations, safety, security, sustainability, constructability, and operability are integrated at various stages of the design process.

A breadth of technical knowledge in several recognized and/or emerging areas of the civil engineering discipline is necessary for understanding the relationship and interaction of different elements in a designed system or environment.

**B: *Design* a system or process to meet desired needs within such realistic constraints as economic, environmental, social, political, ethical, health and safety, constructability, and sustainability. (L5)** The essence of engineering is the iterative process of designing, predicting performance, building, and testing.[4] The National Academy of Engineering recommends that this process be introduced to students from the "earliest stages of the curriculum, including the first year."[4] Fostering creative knowledge in students prepares them to handle a future of increasing complexity that relies on a multidisciplinary approach to problems.[47] The design component in the baccalaureate curriculum should involve both analysis and synthesis.

**E: *Evaluate* the design of a complex system, component, or process and *assess* compliance with customary standards of practice, user's and project's needs, and relevant constraints. (L6)** The post baccalaureate engineering design experience should include opportunities to employ many or all aspects of the design process, including problem definition, project planning, scoping, the design objective, the development of design options, standards, codes, economy, safety, constructability, operability, sustainability, and design evaluation. Experience at this level should include familiarity with interactions between planning, design, construction, and operations and should take into account design life-cycle assessment. The role of peer and senior review and of the design verification process in ensuring successful design should be emphasized to individuals at this level.

# TECHNICAL OUTCOMES

# Outcome 10: Sustainability

## Overview[1]

The 21st-century civil engineer must demonstrate an ability to analyze the sustainability of engineered systems—and of the natural resource base on which they depend—and design accordingly.

ASCE embraced sustainability as an ethical obligation in 1996,[43] and policy statements 418[48] and 517[49] point to the leadership role that civil engineers must play in sustainable development. The 2006 ASCE Summit on the Future of Civil Engineering[1] called for renewed professional commitment to stewardship of natural resources and the environment. Knowledge of the principles of sustainability,[50,51,52] and their expression in engineering practice, is required of all civil engineers.

There are social, economic, and physical[53] aspects of sustainability. The last includes both natural resources and the environment. Technology affects all three, and a broad, integrative understanding is necessary in support of the public interest. Beyond that, special competence is required in the scientific understanding of natural resources and the environment, which are the foundation of all human activity, and the integration of this knowledge into practical designs that support and sustain human development is essential. Vest[54] referred to this as the primary systems problem facing the 21st-century engineer.

The actual life of an engineered work may extend well beyond the design life; and the actual outcomes may be more comprehensive than the initial design intentions. The burden of the engineer is to address sustainability in this longer and wider framework.

Individual projects make separate claims on the collective future; ultimately they cannot be considered in isolation. A commitment to sustainable engineering implies a commitment by the profession to the resolution of the cumulative effects of individual projects. Ignoring cumulative effects can lead to overall failure. This concern must be expressed by the profession generally, and affect its interaction with civil society.

**B: *Apply* the principles of sustainability to the design of traditional and emergent engineering systems.** (L3) Implied is mastery of the scientific understanding of natural resources and the environment and the ethical obligation to relate these sustainably to the public interest. This mastery must rest on a wide educational base,[142] supporting two-way communication with the service population about the desirability of sustainability and its scientific and technical possibilities.

1. See Appendix L for additional ideas and information about sustainability.

**E: *Analyze* systems of engineered works, whether traditional or emergent, for sustainable performance.** (L4) Analysis assumes a scientific, systems-level integration and evaluation of social, economic, and physical factors—the three aspects of sustainability. Achievement at this level requires the "B" achievement described above to be advanced in practice to the analysis level through structured experience and in synergy with other real works, built or planned. Successful progression of cognitive development in this experiential phase must be demonstrable.

# TECHNICAL OUTCOMES

# Outcome 11: Contemporary Issues and Historical Perspectives

## Overview

To be effective, professional civil engineers should draw upon their broad education to analyze the impacts of historical and contemporary issues on engineering and analyze the impact of engineering on the world. The engineering design cycle illustrates the dual nature of this outcome. In defining, formulating, and solving an engineering problem, engineers must consider the impacts of historical events and contemporary issues.

Knowledge of history and heritage helps communicate the importance of the civil engineering profession to society. A noted engineering historian stated that "the value of engineering history goes beyond its being part of the liberal education of an engineer. Engineering history is useful, if not essential, to understanding the nature of engineering; it also assists in the practice of the profession. We gain perspective across fields of engineering by knowing their various and interrelated histories. A historical perspective assists engineers in identifying failure modes and catching errors in logic and design. Engineering history, in short, is engineering as well as history."[55]

Examples of contemporary issues that could impact engineering include the multicultural globalization of engineering practice; raising the quality of life around the world; the importance of sustainability; the growing diversity of society; and the technical, environmental, societal, political, legal, aesthetic, economic, and financial implications of engineering projects. When generating and comparing alternatives and assessing performance, engineers must also consider the impact that engineering solutions have on the economy, environment, political landscape, and society.

**B: Drawing upon a broad education, *explain* the impact of historical and contemporary issues on the identification, formulation, and solution of engineering problems and *explain* the impact of engineering solutions on the economy, environment, political landscape, and society.** (L3) At the undergraduate level, engineers must apply their broad education, derived upon in part from the humanities and social sciences breadth, to the solution of engineering problems. Engineering does not occur in a vacuum, and engineers must be able to both explain the impact of historical and contemporary issues on engineering and explain the impact of engineering on the world. Since contemporary issues and historical perspectives are essentially a part of the engineering design process, most engineering design courses, especially capstone design experiences, include application of contemporary issues and historical perspectives. Examples of other opportunities to incorporate historical perspectives in an undergraduate civil engineering program include providing historical vignettes on scientists and engineers who developed key equations, background and field trips to historical landmarks, and written and oral presentations on historical perspectives of current or future projects.

**E: *Analyze* the impact of historical and contemporary issues on the identification, formulation, and solution of engineering problems and *analyze* the impact of engineering solutions on the economy, environment, political landscape, and society.** (L4) The formal education process sets the stage for future development of the skills required to incorporate historical perspectives and contemporary issues into engineering. In practice, most projects and engineering design involve varying degrees of integration with historical and contemporary issues. The development of the required analytical skills should come from lifelong learning, mentorship from senior engineers, and from practical experience.

# TECHNICAL OUTCOMES

# Outcome 12: Risk and Uncertainty

## Overview

In the past, the teaching of statistics has usually been performed outside of the civil engineering department and rarely integrated into department coursework. Therefore, the student is not significantly exposed to the fundamental concepts of risk/uncertainty within the engineering courses. Civil engineers must deal with real-world uncertainty in design and planning in which public safety is among the top priorities of any design. These uncertainties are unavoidable in any engineering design and in decision-making. They can be data-based or knowledge-based. The engineer should be able to recognize and quantify those uncertainties as part of the design process, apply probability and statistics (P&S) to quantify the risk of failure for well-defined engineering designs, and determine appropriate safety factor(s) to minimize the risk to public safety.

The fundamentals of P&S, in combination with other engineering mathematics and sciences (for example, physics, mechanics, and chemistry), are essential for modeling and analysis of uncertainty, including the identification of all major uncertainties, determining their significance, and applying P&S to assess probability of failure and risk. The fundamentals of P&S should be emphasized in the civil engineering departments, where the practical significance and importance of P&S in engineering can be properly and adequately addressed. The student should be exposed to P&S concepts and their application as early as possible within the civil engineering department curriculum so as to enhance knowledge and understanding of the relationship between P&S and civil engineering applications.

**B: *Apply* the principles of probability and statistics to *solve* problems containing uncertainties.** (L3) A basic understanding of risk and uncertainty principles must be incorporated into the civil engineering department courses. The application of risk/uncertainty within the baccalaureate program should pertain to the design of engineered components within fundamental engineering coursework, in order to quantify the capability of an engineered design component. Individuals should develop sufficient understanding of P&S so they can model problems under uncertainty and interpret the quality of data and its uncertainty obtained from various engineering tools.

**E: *Analyze* the loading and capacity, and the effects of their respective uncertainties, for a well-defined design and *illustrate* the underlying probability of failure (or nonperformance) for a**

**specified failure mode.** (L4) The engineer must be able to review all relevant information and quantify the underlying risk/uncertainty of individual engineered components within a well-defined design or system and determine their significance to and/or synergistic effect on the overall design. The engineer must be able to apply P&S tools to the overall design or system to determine the probability of failure for expected failure modes and be able to illustrate and communicate this information to decision makers and the non-technical community. This knowledge shall form the quantitative basis for selecting the proper safety factor(s) within the design or system.

# TECHNICAL OUTCOMES

# Outcome 13: Project Management

## Overview

Management is a field that touches every individual to some extent—from home to work to the community. In simplest terms, management can be defined as "the act, art, or manner of managing or handling, controlling, directing."[56] Engineering management is the act of managing the engineering relationships among the management tasks related to staffing, organizing, planning, financing, and the human element in production, research, engineering, and service organizations. Engineering managers must understand and integrate organizational, technical, external, and behavioral variables and constraints in order to accomplish predetermined tasks and goals. According to the Project Management Institute, "Project management is the application of knowledge, skills, tools, and techniques to project activities to meet project requirements. Project management is accomplished through the application and integration of the project management processes of initiating, planning, executing, monitoring and controlling, and closing."[57] Project management is best understood within a body of knowledge that is generally recognized as good practice.[58]

B: ***Develop*** **solutions to well-defined project management problems.** (L3) The formal education process has the potential to make a significant impact on teaching project management principles and developing effective managers, including the ability to work alongside and report to people from other cultures. Project management principles include those actions necessary to initiate, plan, execute, monitor and control, and close a project. Examples of curricular project management opportunities in the undergraduate program are design teams for course assignments, capstone design projects, and undergraduate research. Co- and extracurricular project management opportunities include cooperative education assignments, student organization projects, and student-based community service projects.

E: ***Formulate*** **documents to be incorporated into the project plan.** (L4) At a professional level, a civil engineer should be capable of analyzing project management and formulating effective strategies within a phase or subproject of a much larger project or program within the context of the five process groups (or phases) of initiating, planning, executing, monitoring and controlling, and closing. Even at the most basic level of project management, one must be able to coordinate and communicate with other engineers, other disciplines and professionals, clients, and other nontechnical people. In addition to knowledge competencies related to these process groups (or phases), the professional should be able to analyze a situation involving one or more of the following project knowledge areas: integration management, scope management, time management, cost management, quality management, human resources management, communication management, risk management, and procurement management. In addition, every situation should be analyzed with respect to relevant professional responsibility and ethical standards.

# TECHNICAL OUTCOMES

# Outcome 14: Breadth in Civil Engineering Areas

## Overview

The ability to identify engineering problems, formulate alternatives, and recommend feasible solutions is a critically important aspect of the professional responsibilities of a civil engineer. Civil engineering is an inherently broad field encompassing a wide array of technical areas that contribute to infrastructure, public health, and safety. Most civil engineering problems draw upon ideas, concepts, and principles from across the discipline. Thus, professional civil engineers must possess technical breadth and strong problem-solving ability in multiple technical areas of the civil engineering discipline. Traditional technical areas appropriate to civil engineering include construction engineering, environmental engineering, geotechnical engineering, surveying, structural engineering, transportation engineering, and water resources engineering. Nontraditional areas may include engineering and science knowledge areas appropriate to an interdisciplinary approach to the solution of civil engineering problems.

Knowledge and breadth of coverage in at least four technical areas appropriate to civil engineering is necessary for individuals to solve a variety of civil engineering problems. Possessing this breadth will enable individuals to function on intradisciplinary teams to design civil engineering projects, and allow the individual to integrate knowledge from multiple areas to work in secondary, advanced, and/or emerging technologies.

**B: *Analyze* and solve well-defined engineering problems in at least four technical areas appropriate to civil engineering.** (L4) The breadth in technical areas must be obtained at the undergraduate level and should prepare students for subsequent courses in engineering curricula.

# TECHNICAL OUTCOMES

# Outcome 15: Technical Specialization

## Overview

Advanced technical knowledge and skills beyond that included in the traditional four-year bachelor's degree are essential to attaining the BOK necessary for entry into the professional practice of civil engineering. Advanced technical specialization includes all traditionally defined areas of civil engineering practice, but also includes coherent combinations of these traditional areas—that is, advanced knowledge and skills in the area of general civil engineering are appropriate within the context of advanced specialization. Civil engineering specializations in nontraditional, boundary, or such emerging fields as ecological engineering and nanotechnology are suitable and encouraged.

Many nonengineering degrees and courses have content that would be beneficial to the professional practice of civil engineering. These topics/courses may be combined with other appropriate coursework to fulfill the technical specialization and/or other outcomes through the M/30. However, such nonengineering degrees as the M.B.A., J.D., and M.D. would most likely not, by themselves, fulfill the technical specialization of the BOK.

**B: *Define* key aspects of advanced technical specialization appropriate to civil engineering.** (L1) Before one can specialize one must have a basic level of knowledge about advanced technical specialization—that is, an individual must know what is expected of civil engineers that specialize in a particular area. This level of knowledge may be attained through traditional courses as well as through guest lectures by practitioners who practice in the area of interest.

**M/30: *Design* a complex system or process or *create* new knowledge or technologies in a traditional or emerging advanced specialized technical area appropriate to civil engineering.** (L5) In recognition of the ever-advancing profession of civil engineering, advanced technical specialization areas appropriate to civil engineering are, by necessity, open and encompassing of the future needs of the profession. Additionally, discovery and creation of new technologies and knowledge are equally important to the profession's future. Regardless of the specific path towards attainment of technical specialization, tangible relation to the professional practice of civil engineering is required. Individuals are expected to, within their technical area of specialization, synthesize a design, research and develop new methods or tools, and/or discover or create new knowledge or technologies.

**E: *Evaluate* the design of a complex system or process, or *evaluate* the validity of newly created knowledge or technologies in a traditional or emerging advanced specialized technical area appropriate to civil engineering. (L6)** The prelicensure experience should include opportunities to practice—under appropriate guidance and mentorship—civil engineering within the technical area of specialization. The role of practitioner mentorship and review is critical in terms of validating the individual's ability to evaluate, compare and contrast, and validate multiple options within the specific advanced technical area of specialization.

# PROFESSIONAL OUTCOMES

# Outcome 16: Communication

## Overview

Means of communication include listening, observing, reading, speaking, writing, and graphics. The civil engineer must communicate effectively with technical and nontechnical individuals and audiences in a variety of settings. Use of these means of communication by civil engineers requires an understanding of communication within professional practice. Fundamentals of communication should be acquired during formal education. Prelicensure experience should build on these fundamentals to solidify the civil engineer's communication skills.

Within the scope of their practice civil engineers prepare and/or use calculations, spreadsheets, equations, computer models, graphics, and drawings—all of which are integral to a typically complex analysis and design process. Implementation of the results of this sophisticated work requires that civil engineers communicate the essence of their findings and recommendations.

Civil engineers should be acquainted with the tools used to draft their designs. The ability to draw sketches by hand and via computer-aided drafting and design (CADD) software is important in the professional practice of civil engineers. Virtual communication, defined as communication created, simulated, or carried on by means of a computer or other network,[59] is common in engineering practice. Accordingly, civil engineers must be able to use various means of communication in the virtual environment.

**B: *Organize* and *deliver* effective verbal, written, virtual, and graphical communications.** (L4) The undergraduate experience provides many and varied opportunities to present and apply communication fundamentals. Communication can be taught and learned across the curriculum—that is, over all years of formal education and in most courses.

Given the many and varied communication means, communication fundamentals and application can be woven into mathematics, science, and technical and professional practice courses as well as into humanities and social science courses. Examples include having students create graphics to explain complex systems or processes, write detailed laboratory reports for technical audiences and executive summaries for nontechnical audiences, research a topic and write a documented report, and make team presentations in capstone design courses. Such co- and extracurricular activities as cooperative education and active participation in campus organizations offer opportunities to communicate using various means in a variety of situations.

**E: *Plan, compose,* and *integrate* the verbal, written, virtual, and graphical communication of a project to technical and nontechnical audiences.** (L5) Engineering practice provides numerous "real-world" opportunities to apply communication knowledge and skills. The engineer should seek out—or be encouraged to take on—tasks and functions that involve ever more challenging communication. Examples of communication opportunities typically available during the prelicensure period are helping to draft a memorandum or report, using CADD, giving an internal presentation, speaking at local schools, serving on professional society committees, and making a presentation at a conference and publishing the results.

# PROFESSIONAL OUTCOMES

# Outcome 17: Public Policy

## Overview[1]

Public policy is the articulation of the nation's, state's, or municipality's goals and values. Thomas Dye provides a concise statement defining public policy as "whatever governments choose to do or not to do."[60] Since civil engineering is often referred to as a people-serving profession, and the people are the public, civil engineers are inherently part of the public policy process. Whether publicly or privately owned, the civil engineering built environment directly affects the daily lives of people. Civil engineers naturally practice their profession by following standards, specifications, and related guidelines as set forth in public policy documents. Therefore, civil engineers should be exposed to the process of making public policy.

Civil engineers need to have an understanding of public policy and how decision makers in government utilize technical, scientific, and economic information when devising or evaluating public policy. Civil engineers are most involved in public policy in the political process, laws/regulations, funding mechanisms, public education, government-business interaction, and the public service responsibility of professionals. Civil engineering systems have a broad societal context because the core economics of the profession's work is based on public funds, and the use of the built environment is generally available for public use and consumption. Effective integration of civil engineers into the public policy infrastructure is critical to the overall success of society.

The range of public policy issues, processes, and implementation begins with an awareness and understanding of public policy procedures, progresses via systematic evaluation of potential outcomes from public policy decisions, and generally culminates in policy designs and tools to guide future decisions.

**B: *Discuss* and *explain* key concepts and processes involved in public policy.** (L2) Individuals can demonstrate the comprehension of public policy process via such mechanisms as the use of civil engineering standards and regulations in design projects; integration/discussion of local, state, or national civil engineering projects through the curriculum; engagement in public service opportunities; and substantive participation in professional society's activities.

**E: *Apply* public policy process techniques to simple public policy problems related to civil engineering works.** (L3) Individuals can demonstrate the application of public policy techniques via participation that includes active engagement in professional

1. See Appendix N for additional ideas and information about public policy.

societies, the use and development of respective standards of practice, the preparation or review of design/ construction specifications, and economic analysis of alternative design works.

# PROFESSIONAL OUTCOMES

# Outcome 18: Business and Public Administration

## Overview

The professional civil engineer who functions in the business world requires an understanding of business fundamentals. Important business fundamentals topics as typically applied in the private sector include legal forms of ownership, organizational structure and design, income statements, balance sheets, decision (engineering) economics, finance, marketing and sales, billable time, overhead, asset management, profit, and business ethics. The engineer may need a substantially greater amount of business knowledge if he or she plans to work outside the country in the global business environment.

The professional civil engineer who functions within the public sector requires an understanding of public administration fundamentals. Essential public administration fundamentals include the political process, laws and regulations, funding mechanisms, public education and involvement, governmental-business interaction, and the public service responsibility of professionals.

**B: *Explain* key concepts and processes used in business and public administration.** (L2) Examples of key concepts include problem identification, denoting the type of personnel and/or organization required to solve the problem, implications of the current laws and regulations, and responsibility to the public and/or client. Examples of the problem-solving process include identifying applicable technologies to address the issue, developing budgets and project schedules, understanding funding mechanisms, and business ethics.

**E: *Apply* business and public administration concepts and processes.** (L3) The individual should be able to set personnel requirements and time allocations to accomplish specific tasks, determine sufficient funding for projects, and understand the basic principles of a balance sheet and billing/payment requirements. The individual should also be able to determine the necessary government/business interaction and contractual requirements for projects.

# PROFESSIONAL OUTCOMES

# Outcome 19: Globalization

## Overview[1]

The world is increasingly interconnected. Countries and their social, constructed, and natural environments demonstrate emerging interdependencies that must be considered in planning and selecting projects. Immediate access to information is everywhere, and in many respects geographic proximity is becoming less important to the success of a project.[61] Engineers will need to deal with ever-increasing globalization; find ways to prosper within an integrated international environment; and meet challenges that cross cultural, language, legal, and political boundaries while respecting critical cultural constraints and differences.[1,62]

The 21st-century civil engineer must address three distinct global topics:[63,50] the globalization process, global issues, and global professionalism. Examples of the globalization process are globalization's effect on infrastructure revitalization within the U.S.; the dependency of economic wealth on the variety, reliability, and service of physical infrastructure systems; and cost and governmental issues such as taxation and subsidy differences across jurisdictions.

1. See Appendix M for additional ideas and information about globalization.

Global issue examples include the international scale of such extreme and long-term environmental events as natural disasters, global climate change, and their impacts on the natural, built, and social environments; meeting a world health standard; and developing acceptable international standards for both large and developing countries.

Examples of global professionalism issues are individuals, businesses, and the profession becoming effective in multicultural practice; the challenge of practicing ethically in a global environment; and barriers to professional licensure, contractor licenses/permits, and foreign corporations.

**B: *Organize, formulate,* and *solve* engineering problems within a global context.** (L3) An individual should be able to apply the knowledge gained through working on course projects and participation in other activities to organize, formulate, and solve problems involving global issues. For example, apply the fundamental knowledge of handling global issues in specific applications and scenarios.

**E: *Analyze* engineering works and services in order to function at a basic level in a global context.** (L4) Analysis includes the impact of the globalization process, global issues, and global professionalism. Ways that individuals can achieve this include industry interaction and professional society participation.

# PROFESSIONAL OUTCOMES

# Outcome 20: Leadership

## Overview

In a broad sense leadership is developing and engaging others in a common vision, clearly planning and organizing resources, developing and maintaining trust, sharing perspectives, inspiring creativity, heightening motivation, and being sensitive to competing needs. Leadership is the art and science of influencing others toward accomplishing common goals and does not necessarily require a formal role or position within a group. Engineers must be willing to lead when confronted with professional and/or ethical issues. More often "employers [are] calling for graduates who are not merely expert in design and analysis but who possess the leadership skills to apply their technical expertise and to capitalize on emerging construction and information technologies, management models, and organizational structures."[64] Many also argue that "an engineer is hired for his or her technical skills, fired for poor people skills, and promoted for leadership and management skills."[65]

Although technical competence and broad managerial skills will remain important, success in engineering will be more a result of leadership in applying that competence and those skills, rather than the competence and skills themselves.[4] The NAE report *The Engineer of 2020: Visions of Engineering in the New Century* states that "engineers must understand the principles of leadership and be able to practice them as their careers advance."[4] Clearly the acquisition of leadership skills and the art of practicing leadership are vital to the future of civil engineering. By the very nature of a profession that requires the attainment of strong analytical and rational decision making skills, engineers are particularly well suited to assume leadership roles.

**B: *Apply* leadership principles to direct the efforts of a small, homogeneous group.** (L3) The best place to start the formal leadership development process is at the undergraduate level.[66] Leadership can be taught and learned. Leadership principles include being technically competent, knowing oneself and seeking self improvement, making sound and timely decisions, setting the example, seeking responsibility and taking responsibility for one's actions, communicating with and developing subordinates both as individuals and as a team, and ensuring that the project is understood, supervised, and accomplished. The formal education process has the potential to make a significant impact on teaching leadership principles and developing leadership attributes.[64]

Qualities and attributes of leaders include:[66,67,68] vision, enthusiasm, industriousness, initiative, competence, commitment, selflessness, integrity, high ethical standards, adaptability, communication skills, discipline, agility, confidence, courage, curiosity, and persistence. Examples of leadership opportunities in the undergraduate program include leadership of design teams, leadership opportunities within capstone designs, and leadership within such organizations as ASCE's student chapters, student competitions, civic organizations, honor societies, athletic teams, student government, and fraternities and sororities.

**E: *Organize* and *direct* the efforts of a group.** (L4) Leadership cannot be solely acquired in a classroom. Leadership development during formal education must be reinforced by extensive practice in real-world settings early in an engineer's career,[66] and leadership development must continue throughout an engineer's career. Senior engineers must mentor junior engineers and provide opportunities for leadership.

# PROFESSIONAL OUTCOMES

# Outcome 21: Teamwork

## Overview

Licensed civil engineers must be able to function as members of a team. This requires understanding team formation and evolution, personality profiles, team dynamics, collaboration among diverse disciplines, problem solving, and time management and being able to foster and integrate diversity of perspectives, knowledge, and experiences.[69]

A civil engineer will work within two different types of teams. The first is intradisciplinary and consists of members from within the civil engineering subdiscipline—for example, a structural engineer working with a geotechnical engineer. The second is multidisciplinary and is a team composed of members of different professions—for example, a civil engineer working with an economist on the financial implications of a project or a civil engineer working with local elected officials on a public planning board. Multidisciplinary also includes a team consisting of members from different engineering subdisciplines—sometimes referred to as a crossdisciplinary team—for example, a civil engineer working with a mechanical engineer.

**B: *Function* effectively as a member of an intradisciplinary team.** (L3) At the undergraduate level, the focus is primarily on working as members of an intradisciplinary team—that is, a team within the civil engineering subdiscipline. Effective team members are usually honest, open-minded, tolerant, diligent, reliable, and considerate. Examples of opportunities for students to work in teams include design projects and laboratory exercises within a course and during a capstone design experience. The development of the ability to function as a member of a team goes beyond the classroom and engineering. Accordingly, students should seek opportunities to work as team members in myriad activities, including student government, civic and service organizations, and employment.

**E: *Function* effectively as a member of a multidisciplinary team.** (L4) Prior to licensing, engineers must be able to effectively function as a member of a multidisciplinary team. Engineers must be able to work with other engineers outside the civil engineering discipline or others outside the engineering profession. In practice, most projects and designs will incorporate other engineering disciplines and/or other professions. For example, civil engineers will often have to work with mechanical engineers for a structural building design, with environmental engineers for a water resources project, or with construction management personnel on many projects. Civil engineers may also work with public planning boards or financial consultants on projects. By the very nature of the profession, civil engineers need to develop and exercise strong teamwork skills.

# PROFESSIONAL OUTCOMES

# Outcome 22: Attitudes

## Overview[1]

Attitudes fundamentally and profoundly affect the success and welfare of projects and the profession. Attitudes are the ways in which one thinks and feels in response to a fact or situation. Attitudes reflect an individual's values and world view and the way he or she perceives, interprets, and approaches surroundings and situations. Attitudes do not exist in a vacuum but are related to some object or situation. While this definition is very broad, this BOK limits its scope to attitudes supportive of the professional practice of civil engineering.

The positive attitudes generally considered to be conducive to the effective professional practice of civil engineering include commitment, confidence, consideration of others, curiosity, entrepreneurship, fairness, high expectations, honesty, integrity, intuition, judgment, optimism, persistence, positiveness, respect, self esteem, sensitivity, thoughtfulness, thoroughness, and tolerance. The list is not exhaustive and some of the attitudes can manifest themselves in negative ways. The BOK includes only positive, constructive expressions of these and other attitudes.

**B: *Explain* attitudes supportive of the professional practice of civil engineering.** (L2) Beginning to develop supportive attitudes during undergraduate education is important. Certainly these attitudes should be modeled by instructors, advisors, mentors, and others concerned with a student's progress toward a degree. Preferably the student will model these attitudes upon graduation, but what is required here is that the student be aware of and explain attitudes supportive of the professional practice of civil engineering.

**E: *Demonstrate* attitudes supportive of the professional practice of civil engineering.** (L3) The licensed engineer must demonstrate attitudes supportive of professional practice. The engineering process requires individuals to work well with others and assume leadership roles in specific areas. Supportive attitudes are essential to the successful accomplishment of these tasks and to many other outcomes related to professional practice.

1. See Appendix O for additional ideas and information about attitudes.

# PROFESSIONAL OUTCOMES

# Outcome 23: Lifelong Learning

## Overview

Given the ever-increasing quantity of technical and nontechnical knowledge required of practicing civil engineers, the ability to engage in lifelong learning is essential. Lifelong learning is defined as the ability to acquire knowledge, understanding, or skill throughout one's life. Knowledge, skills, and experience acquired in undergraduate programs are not sufficient for a career spanning several decades. Civil engineers should engage in lifelong learning through additional formal education; continuing education; professional practice experience; and active involvement in professional societies, community service, coaching, mentoring, and other learning and growth activities.

**B: *Demonstrate* the ability for self-directed learning.** (L3) At the undergraduate level, the focus is first to define lifelong learning and explain why lifelong learning is an essential skill for the successful practice of engineering. Graduates must also describe the skills required for lifelong learning, demonstrate the ability for self-directed learning, and develop their own learning plan. Self-directed learning is a mode of lifelong learning because it is the ability to learn on one's own with the aid of formal education.[70] Independent study projects and open-ended problems that require additional knowledge that is not presented in a formal class setting are examples of ways to provide opportunities for self-directed learning in an undergraduate program. Programs can also assess student work requiring professional goal setting or reflection on the value of lifelong learning. Student participation in professional development activities such as professional society membership, community service, and preparation for the Fundamentals of Engineering exam are also examples of lifelong learning.

**E: *Plan* and *execute* the acquisition of expertise appropriate for professional practice.** (L5) Prior to licensing, engineers must first be able to identify additional knowledge, skills, and attitudes appropriate for professional practice. Engineers must then be able to plan and execute the acquisition of knowledge, skills, and experiences required for professional practice and plan and execute their own professional development program in response to internal and external motivations. Lifelong learning activities include personal and professional development on goal setting, personal time management, delegation, understanding personality types, networking, leadership, appreciating sociopolitical processes, and affecting change. Other types of professional development include career management, increasing knowledge in a specific discipline, contributing to the profession through service on committees in professional organizations, additional formal education, and achieving specialty certification. Mentorship should play a key role in the lifelong learning process. Finally, civil engineers must have the ability to learn how to learn.[71]

# PROFESSIONAL OUTCOMES

# Outcome 24: Professional and Ethical Responsibility

## Overview

Civil engineers in professional practice have a privileged position in society, affording the profession exclusivity in the design of the public's infrastructure. This position requires each of its members to adhere to a doctrine of professionalism and ethical responsibility. This doctrine is set forth in the seven fundamental canons in ASCE's Code of Ethics.[43] The first canon states that civil engineers "...shall hold paramount the safety, health, and welfare of the public...." By meeting this responsibility, which puts the public interest above all else, the profession earns society's trust.

According to the vision for civil engineering in 2025,[1] civil engineers aspire to be "entrusted by society to create a sustainable world and enhance the global quality of life." Therefore, current and future civil engineers, whether employed in public or private organizations or self-employed, will increasingly hold privileged and responsible positions. Although the fundamental canons may detail the appropriate behavior and attitude of the individual, consistent with the privilege of membership in the profession of civil engineering, professional and ethical behavior goes beyond the minimums defined by ethics codes. Depending on the individual's interests and circumstances, professional and ethical activities may include mentoring less experienced personnel, leading or actively participating in professional societies, and involvement in community affairs.

**B: *Analyze* a situation involving multiple conflicting professional and ethical interests to determine an appropriate course of action.** (L4) The undergraduate experience should introduce and illustrate the impact of the civil engineer's work on society and the environment. This naturally leads to the importance of meeting such professional responsibilities as maintaining competency and the need for ethical behavior. The latter can be aided by familiarity with engineering codes of ethics and by identifying professional engineers and licensing boards as additional guidance resources. Going beyond satisfying codes, students should begin to see the opportunities that their profession offers for participation, including leadership in professional societies and community affairs. The preceding teaching and learning can be accomplished across the curriculum, including by example, and in selected co- and extracurricular activities including participation in cooperative education and active involvement in engineering professional societies and campus organizations.

**E: *Justify* a solution to an engineering problem based on professional and ethical standards and *assess* personal professional and ethical development.** (L6) The professional is likely to quickly encounter professional and ethical issues in his/her professional career. In fact, supervisors, coaches, and mentors should offer the professional opportunities to participate in applying pertinent laws and regulations and professional and ethical principles to help define and resolve such issues. The individual should be encouraged to continuously enhance his or her professional and ethical development by becoming actively involved in professional societies and community affairs.

# APPENDIX K

# Humanities and Social Sciences

## Introduction

What is the role of engineers in society, and how is that role changing? The National Academy of Engineering report *The Engineer of 2020*[a] identifies these three visions for the engineering profession:

- By 2020, we aspire to a public that will understand and appreciate the profound impact of the influence of the engineering profession on sociocultural systems, the full spectrum of career opportunities accessible through an engineering education, and the value of an engineering education to engineers working successfully in nonengineering jobs.
- We aspire to a public that will recognize the union of professionalism, technical knowledge, social and historical awareness, and traditions that serve to make engineers competent to address the world's complex and changing challenges.
- We aspire to engineers who will remain well grounded in the basics of mathematics and science, and who will expand their vision of design through solid grounding in the humanities, social sciences, and economics. Emphasis on the creative process will allow more effective leadership in the development and application of next-generation technologies to problems of the future.

The need for humanities and social sciences (H&SS) education for engineers is evident in each of these statements. The humanities include such subjects as art, religion, philosophy, history, and literature while the social sciences include such subjects as economics, political science, sociology, and psychology. Social sciences are often data-driven and quantitative while the humanities typically employ critical and analytic thinking.

Related to the preceding vision is the global vision that emerged from ASCE's 2006 Summit on the Future of Civil Engineering.[b] The vision, which is included in Chapter 2 of this report, describes where the civil engineering profession will strive to be in 2025.

Fulfillment of the civil engineering vision requires professional activity supported on a balanced base of liberal learning. Failure to provide civil engineers with an education founded upon this balanced base will compromise the profession's ability to realize this vision; to recruit and retain the best talent; and to perform effectively as a profession. This concept is broadly shared among other professions, including law, medicine, and architecture.

The ASCE vision asserts important aspirations for civil engineering. A commitment to this vision must be reflected in substantive ways in the civil engineering BOK.

## Liberal Learning in Civil Engineering Education

Liberal learning—defined as learning that frees the mind—is normally founded upon four general areas of education: natural sciences, mathematics, the humanities, and the social sciences. Liberal learning implies free and broad inquiry with intellectual discipline and is foundational for many other established professions. Engineering needs this balanced base of education.

Because liberal learning underpins other professions, an independent description of liberal learning is useful. The following statement was adopted by the Board of Directors of the Association of American Colleges & Universities (AACU) in October 1998:

> *A truly liberal education is one that prepares us to live responsible, productive, and creative lives in a dramatically changing world. It is an education that fosters a well-grounded intellectual resilience, a disposition toward lifelong learning, and an acceptance of responsibility for the ethical consequences of our ideas and actions. Liberal education requires that we understand the foundations of knowledge and inquiry about nature, culture and society; that we master core skills of perception, analysis, and expression; that we cultivate a respect for truth; that we recognize the importance of historical and cultural context; and that we explore connections among formal learning, citizenship, and service to our communities.*

> *We experience the benefits of liberal learning by pursuing intellectual work that is honest, challenging, and significant, and by preparing ourselves to use knowledge and power in responsible ways. Liberal learning is not confined to particular fields of study. What matters in liberal education is substantial content, rigorous methodology, and an active engagement with the societal, ethical, and practical implications of our learning. The spirit and value of liberal learning are equally relevant to all forms of higher education and to all students.*

> *Because liberal learning aims to free us from the constraints of ignorance, sectarianism, and myopia, it prizes curiosity and seeks to expand the boundaries of human knowledge. By its nature, therefore, liberal learning is global and pluralistic. It embraces the diversity of ideas and experiences that characterize the social, natural, and intellectual world. To acknowledge such diversity in all its forms is both an intellectual commitment and a social responsibility, for nothing less will equip us to understand our world and to pursue fruitful lives.*

> *The ability to think, to learn, and to express oneself both rigorously and creatively, the capacity to understand ideas and issues in context, the commitment to live in society, and the yearning for truth are fundamental features of our humanity. In centering education upon these qualities, liberal learning is society's best investment in our shared future.*

The following statement describes, from the above AACU source, the essence of liberal learning. It must be embraced by civil engineering education as part of the process of preparing civil engineers of the future.

> *What matters in liberal education is substantial content, rigorous methodology, and an active engagement*

*with the societal, ethical, and practical implications of our learning.*

Doesn't this describe the goals and aspirations of civil engineering education?

The need for H&SS in civil engineering education is supported by the concepts of liberal learning and the concepts of critical thinking. Civil engineers think about and develop solutions to problems. A civil engineer's thinking must be systematic and guided and informed by analysis and assessment of relevant information. A civil engineer's thinking must not be arbitrary, biased, lacking in context, or poorly substantiated. A critical thinker:[c]

- "raises vital questions and problems, formulating them clearly and precisely;
- gathers and assesses relevant information, using abstract ideas to interpret it effectively, comes to well-reasoned conclusions and solutions, testing them against relevant criteria and standards;
- thinks open-mindedly in consideration of alternative solutions, recognizing and assessing, as need be, their assumptions, implications, and practical consequences; and
- communicates effectively with others in figuring out solutions to complex problems."

For civil engineers educated exclusively in areas of mathematics and science, the most prominent questions are likely to be mathematical and scientific questions. Alternatively, a civil engineer whose education includes H&SS will bring more to the critical thinking process. A broadly educated engineer is likely to recognize the impact of the engineering decisions not only upon the more narrowly framed mathematics, science, and technical questions but upon the more broadly framed questions informed by H&SS.

## A Balanced Body of Knowledge

Figure K-1 attempts to capture the central idea of a broad education graphically by showing technical and professional education and performance supported by four foundational legs (mathematics, natural sciences, humanities, social sciences). Together these broadly capture the established dimensions of higher education.

The 20$^{th}$ century witnessed a major expansion in the mathematics and science "legs" that support civil engineering. The continuing importance of this is emphasized by the inclusion of four separate outcomes in BOK2 and the strong reliance of other outcomes on this mathematics science foundation.

An absence of explicit support legs for the humanities and social sciences would be consistent with the classical impression of an engineer well grounded only in technical issues. This is an unfortunate historical stereotype—one that the profession large rejects today, and aspires to move beyond. Accordingly, separate support legs for humanities and for social sciences are included in the BOK2. All four legs are essential in supporting the vision of the civil engineering profession.

Relative to these legs, Vest,[d,e] identifies two "pivotal" developments in engineering education since World War II: the development of the science base of "engineering science"; and the incorporation of the H&SS in support of the "21st-century view of engineering systems, which surely are not based solely on physics and chemistry." Note the increasing reliance on the humanities leg and social sciences leg and the obligation to develop these within the profession broadly, as a matter of basic professional competence.

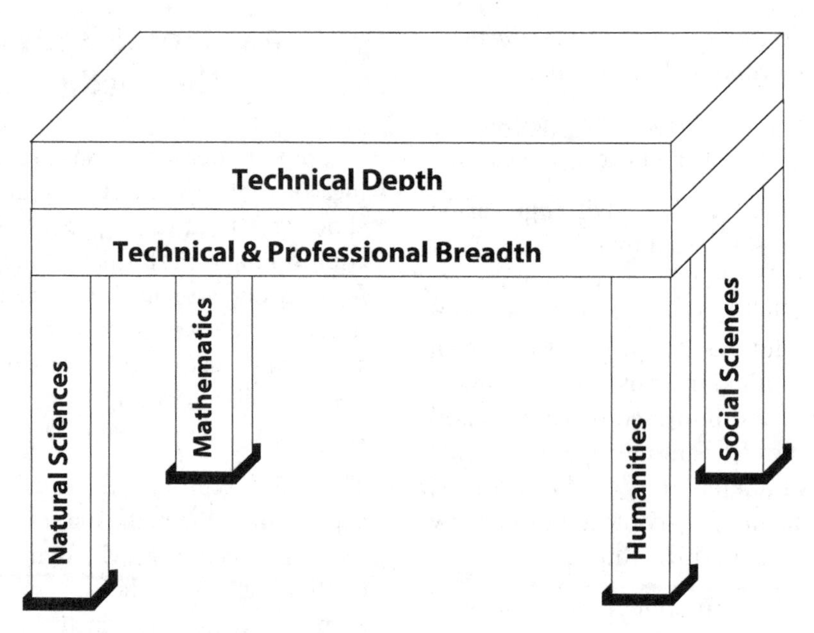


Figure K-1. The future technical and professional practice education of civil engineers is supported on four foundational legs.

## Foundational Outcomes in the Body of Knowledge

In order to recognize the importance of humanities and social sciences in the education of future civil engineers, two new outcomes—one for humanities and one for the social sciences—have been included in the civil engineering BOK. There is considerable freedom for educators to determine how these outcomes may be fulfilled through contributions from various academic departments and disciplines. Unlike the basic sciences outcomes (chemistry, physics, and natural sciences), particular humanity or social science fields are not specified. This freedom permits each program to devise requirements consistent with their university and department missions.

## Cited Sources


a) National Academy of Engineering. 2004. *The Engineer of 2020: Visions of Engineering in the New Century.* The National Academy of Sciences, Washington, D.C. (http://www.nae.edu)

b) ASCE Task Committee to Plan a Summit on the Future of the Civil Engineering Profession. 2007. *The Vision for Civil Engineering in 2025,* ASCE, Reston, VA. (A PDF version is available, at no cost, from http://www.asce.org./Vision2025.pdf)

c) Foundation for Critical Thinking, "The Critical Thinking Community." (http://www.criticalthinking.org/)

d) Vest, C., "Educating Engineers for 2020 and Beyond," in National Academy of Engineering. 2004. *The Engineer of 2020: Visions of Engineering in the New Century.* The National Academy of Sciences, Washington, D.C. (http://www.nae.edu)

e) Vest, C. 2006. "Educating Engineers for 2020 and Beyond," *The Bridge,* Summer, p. 40.

# APPENDIX L

# Sustainability

*The 21st-century civil engineer must demonstrate an ability to analyze the sustainability of engineered systems—and of the natural resource base on which they depend—and design accordingly.*

## Overview

Civil engineering developed historically with a distinctive focus on civilian infrastructure and the technological support of civil society generally. The profession has continued to affirm this mission throughout the 20th century and into the 21st century. Necessarily, technology continues to evolve, and problems mirror society in their increasing scale and complexity. The globalization of civil society has brought a parallel globalization of civil engineering concerns and practice.[a,b] A primary dimension of those concerns and practice is *sustainability.*

Unquestionably, global scenarios involve the natural resource base that sustains civil society and the natural and the built environments. Example concerns include the depletion of fossil resources; the management of new energy sources, including the nuclear fuel cycle; the bioengineering of fuel, food, and drugs; the maintenance of agricultural productivity; the increasing exploitation of the oceans; the human right to water; nuclear chemistry; and more. Anthropogenic influences are clearly visible in the global ecosystem: species extinction, exhaustion of depleted resources, geopolitical conflict over ownership of renewable resources, and degradation of the planetary commons (for example, the atmosphere, oceans, and rivers). Civil engineering cannot by itself "solve" these problems; yet it must embrace a proactive, professional stance and contribute distinctive competence toward their resolution.

## Civil Engineering and the Sustainability Commitment

ASCE adopted this definition in November 1996:[c]

> *Sustainable development is the challenge of meeting human needs for natural resources, industrial products, energy, food, transportation, shelter, and effective waste management while conserving and protecting environmental quality and the natural resource base essential for future development.*

That this expresses an ethical obligation on the part of the profession has been recognized since 1996 in the ASCE Code of Ethics.[d] Fundamental Canon 1 asserts that

> *Engineers shall hold paramount the safety, health, and welfare of the public and shall strive to comply with the principles of sustainable development in the performance of their professional duties.*

A comparable ethics statement was adopted in 2006 by NSPE;[e]

> *Engineers shall strive to adhere to the principles of sustainable development in order to protect the environment for future generations.*

and a footnote adopted verbatim the ASCE definition of sustainable development.

ASCE Policy Statement 418 (October 2004) affirmed the role of the profession in addressing and securing sustainability:[f]

> *The American Society of Civil Engineers (ASCE) recognizes the leadership role of engineers in sustainable development, and their responsibility to provide quality and innovation in addressing the challenges of sustainability.*

In that document, ASCE committed to implementation strategies, including:

- Promote broad understanding of political, economic, social, and technical issues and processes as related to sustainable development.
- Advance the skills, knowledge, and information to facilitate a sustainable future, including habitats, natural systems, system flows, and the effects of all phases of the life cycle of projects on the ecosystem.
- Promote performance-based standards and guidelines as bases for voluntary actions and for regulations in sustainable development for new and existing infrastructure.

In June 2002, the "Dialog on the Engineers' Role in Sustainable Development – Johannesburg and Beyond"[g] committed the several institutional signatories (AAES, AIChE, ASME, NAE, NSPE) to the declaration:

> *Creating a sustainable world that provides a safe, secure, healthy life for all peoples is a priority for the US engineering community. ... Engineers must deliver solutions that are technically viable, commercially feasible, and environmentally and socially sustainable.*

> *In July 2006, ASCE endorsed the millennium development goals in Policy Statement 517:*[h]

> *The American Society of Civil Engineers (ASCE) supports the internationally agreed development goals contained in the Millennium Declaration as they apply to improving the quality of people's lives around the world through science and engineering. ASCE works in collaboration with other domestic and international organizations to engage engineers in addressing the needs of the poor through capacity building and the development of sustainable and appropriate solutions to poverty.*

Partly in response, the ASCE Committee on Sustainability published *Sustainable Engineering Practice: An Introduction*[i] in 2004. This report

> *...is intended to be a 'primer' on sustainability that ... can inspire and encourage engineers to pursue and integrate sustainable engineering into their work...*

As a primer, this is a gathering of the practical state of the art at the time of its publication. A great deal of practical material is assembled therein. This follows in the path of the earlier ASCE/UNESCO monograph *Sustainability Criteria for Water Resource Systems.*[j]

The notion of a necessary engineering response to sustainability is pervasive in the NAE reports on the engineer of 2020.[k,l] Therein Vest[m] cites sustainability as the top systems integration problem facing engineering today. Reflecting these and other current analyses, ASCE published its *Vision for Civil Engineering in 2025*[n] which opens with:

> *"Entrusted by society to create a sustainable world ... "*

## Sustainability and the Body of Knowledge

The initial BOK[o] was issued in 2004. Sustainability is not represented explicitly in any of the 15 outcomes. Despite the accelerating discussion and professional commitment, the word "sustainability" occurs only twice in the commentaries accompanying outcomes 1 and 3 on page 25 of the first edition of the BOK report.

In this, the second edition of the BOK report, sustainability is incorporated as a new, independent outcome, achievement of which is recommended for entry into the profession. The explanation to outcome 10, (sustainability) begins with this statement:

> *The 21st-century civil engineer must demonstrate an ability to analyze the sustainability of engineered systems—and of the natural resource base on which they depend—and design accordingly.*

This is the natural and necessary complement to the established policy statements about sustainability and to the large volume of practical activity already under way in the profession. Recognizing this outcome in the BOK2 legitimizes the basic preparation of young engineers in sustainability, encourages its scholarly and professional evolution, and devotes the profession at large to lifelong learning and professional practice in pursuit of sustainable outcomes.

Accordingly, three definitions are needed: sustainability, sustainable engineering, and sustainable development. These are all based herein on the extant ASCE definition of sustainable development, quoted above. The other two definitions:

> *Sustainability is the ability to meet human needs for natural resources, industrial products, energy, food, transportation, shelter, and effective waste management while conserving and protecting environmental quality and the natural resource base essential for the future.*
>
> *Sustainable engineering meets these human needs.*

## Interdisciplinary, Distinctive Competence, Scope

There are social, economic, and physical aspects of sustainability. Each affects, and is affected by, technology, natural resources, and the environment. A broad, integrative understanding of all of these aspects is necessary. Beyond that, special competence is required in the scientific understanding of natural resources and the environment, which are the foundation of all human activity; and the integration of this knowledge into practical designs that support and sustain human development.

Special technical competence is required in three areas:

- Sustaining the availability and productivity of natural resources, the ultimate base of civil society;
- Sustaining civil infrastructure, the engineered environment; and
- Sustaining the environment generally, the human habitat.

This technical competence must rest on a broad scientific base, including biological and chemical phenomena and be sufficient to support relevant emerging technologies, creative design, and natural resource constraints. There is no substitute for science in an engineer's competence. And beyond that, equally essential is the ability to innovate—to recognize and solve problems with judicious technical approaches.

Other critical dimensions of sustainability are the economic, social, and political aspects of civil life. Competence must rest on a proper foundation here, too—in the humanities, supporting human aspirations and their expression; and in the social sciences, supporting effective use of political and economic means in assessing and meeting needs. The breadth of the sustainability challenge was captured in a recent article "Sustainability: the Ultimate Liberal Art."[p]

Beyond these claims on the base, sustainability makes claims on the research frontier. No one would assert that the sustainability "problem" is a closed one, solutions lying in established forms. An aggressive search for the knowledge necessary to make advances in this area is clearly needed.

## The Rationale behind the Sustainability Rubric

Six levels of achievement are defined for each BOK2 outcome. The different levels generally build on a precollege preparation. For sustainability, the rationale for each level further explains its meaning.

**Level 1—Knowledge:** ***Define*** key aspects of sustainability relative to engineering phenomena, society at large, and its dependence on natural resources, and relative to the ethical obligation of the professional engineer. *Rationale:* Proactive integration of diverse considerations is implied at the point where an engineering solution is proposed and evaluated. Implied is an ability to conceive of the full life-cycle of an engineering project and a comprehensive set of outcomes, including effects on the environment, the natural resource base, the conditions at project termination, and the appropriateness of the project itself and how it serves the public interest.

**Level 2—Comprehension:** ***Explain*** key properties of sustainability, and their scientific bases, as they pertain to engineered works and services. *Rationale:* This is the natural extension of level 1. A blend of theory and experiment is likely in applying ideas to engineered systems. A scientific explanation is necessary, especially relative to natural resources and to the natural and built environment, where established scientific descriptions are available

**Level 3—Application:** ***Apply*** the principles of sustainability to the design of traditional and emergent engineering systems. *Rationale:* This is the natural extension of level 2. Graduates must be capable of applying ideas to real engineering works and of utilizing general information available within the profession.

**Level 4—Analysis:** *Analyze* systems of engineered works, whether traditional or emergent, for sustainable performance. *Rationale:* This is a systems-level integration of cumulative and synergistic effects of works with respect the sustainability of the composite outcome. Implied is the ability to propose and compare alternatives in an analytic framework.

**Level 5—Synthesis:** ***Design*** a complex system, process, or project to perform sustainably. ***Develop*** new, more sustainable technology. ***Create*** new knowledge or forms of analysis in areas where scientific knowledge limits sustainable design. *Rationale:* This is either professional-strength design or research. The latter can have varying amounts of scientific overlap.

**Level 6—Evaluation:** *Evaluate* the sustainability of complex systems, whether proposed or existing. *Rational:* This refers to the ability to inspire and evaluate the work of teams engaged synergistically. Included is the ability to quantify the value of research in sustainable engineering.

The higher levels begin to address the abilities of the profession as a group, presumably characterized by a broad baseline of competence, the presence of accomplished specialists, a demonstration of that competence sufficient to earn the public trust, and a collective commitment to sustainability.

## Cited Sources


a) Jha, M.K. and D. Lynch. 2007. "Role of Globalization and Sustainable Engineering Practice in the Future Civil Engineering Education," in *Sustainable Development and Planning, III,* A Kungolos, C. A. Brebbia, E. Beriatos (eds), Vol. 2, 641–650, 2007.

b) Lynch D., W. Kelly, M. K. Jha, and R. Harichandran. 2007. "Implementing Sustainably in the Engineering Curriculum: Realizing the ASCE Body of Knowledge," proceedings of the ASEE Annual Conference, Honolulu, HI, June.

c) ASCE. 1996. The definition of sustainable development adopted by the ASCE Board of Direction. https://www.asce.org/inside/codeofethics.cfm (see footnote therein for details)

d) ASCE. 1996. "Code of Ethics." https://www.asce.org/inside/codeofethics.cfm

e) NSPE. 2006. "Code of Ethics." http://www.nspe.org/ethics/eh1-codepage.asp

f) ASCE. 2004. Policy 418, "The Role of the Civil Engineer in Sustainable Development," adopted by the Board of Direction, October 19 2004. http://www.asce.org/pressroom/news/policy_details.cfm?hdlid=60

g) National Academy of Engineering. 2002. "Dialogue on the Engineer's Role in Sustainable Development—Johannesburg and Beyond." June 24, 2002. U.S. National Academies; statement endorsed by AAES, AIChE, ASME, NAE, and NSPE.

h) ASCE. 2006: "Millennium Development Goals," Policy Statement 517. http://www.asce.org/pressroom/news/policy_details.cfm?hdlid=514

i) ASCE Committee on Sustainability. 2004. *Sustainable Engineering Practice: An Introduction.*

j) Loucks, D. P. and J. S. Gladwell. 1999. *Sustainability Criteria for Water Resource Systems,* Cambridge University Press/UNESCO, 1999, 139 pp.

k) National Academy of Engineering. 2004. *The Engineer of 2020: Visions of Engineering in the New Century.*

l) National Academy of Engineering. 2005. *Educating the Engineer of 2020: Adapting Engineering Education to the New Century.*

m) Vest, C. M. 2005. "Educating Engineers for 2020 and Beyond," in Educating the Engineer of 2020, NAE 2005 (op. cit.); pp. 160–169.

n) ASCE. 2006. *The Vision for Civil Engineering in 2025. Report of the Summit on The Future of Civil Engineering.*

o) ASCE Body of Knowledge Committee. 2004. *Civil Engineering Body of Knowledge for the 21st Century: Preparing the Civil Engineer for the Future.*

p) Rhodes, F. T. 2006. "Sustainability: The Ultimate Liberal Art." *The Chronicle of Higher Education*, October 20.

q) Lynch, D. R. 2007. "Sustainability Background Information." http://www-nml.dartmouth.edu/Publications/internal_reports/NML-06-Sustain/

Bibliographic background is available.¶

# APPENDIX M

# Globalization

## Introduction

In recent years globalization has been at the core of many studies reported by the National Academy of Engineering (NAE), the National Science Foundation (NSF), and the American Society for Engineering Education (ASEE). The civil engineering profession deals with issues that may have global impact, including the outsourcing of engineering services, the design and construction of civil engineering infrastructure, and creating a world health standard by providing adequate sanitation facilities and drinking water. In the wake of rapid advancement in information technology, as well as increasing diversification of society and the pressing need for understanding global issues, tomorrow's civil engineers must be prepared to handle global aspects of civil engineering practice.

The world is increasingly interconnected. Countries and their social, constructed, and natural environments demonstrate emerging interdependencies that must be considered in planning and selecting projects. Information is increasingly readily obtained and, in many respects, geographic proximity is becoming less important to the success of a project. Engineers will need to deal with ever-increasing globalization and find ways to prosper within an integrated international environment and meet challenges that cross cultural, language, legal, and political boundaries while respecting critical cultural constraints and differences.

Outcome 19 (globalization) addresses three distinct topics: the globalization process, global issues, and global professionalism. Practitioners network globally. Problems are local, culturally mediated, and diverse. Although the underlying principles are universal, "one size does not fit all" when seeking solutions. The global professional must be able to deliver effective solutions in diverse cultures, with diverse cooperation. A license to practice in, for example, Wyoming, must be respected beyond state and U.S. borders. Its holder must be capable of, and accepted for, work beyond the local jurisdiction. Specifically, the 21$^{st}$-century civil engineer must demonstrate the impact of globalization on the following four areas.

## Professional Practice

What will be the effect of ever-increasing globalization on the practice of civil engineering in the 21$^{st}$ century? Possible issues include:

- Individuals, businesses, and the profession becoming effective in multicultural practice
- The challenge of practicing ethically in a global environment

- Civil engineers of the 21st century adapting to fully participate in the global economy
- Working in a borderless, diverse culture
- Bringing innovation back from overseas
- Barriers to professional licensure, contractor licenses/permits, and foreign corporations

## Infrastructure

What will be the civil engineer's role in creating and maintaining the physical infrastructure throughout the world in the 21st century? Possible issues include:

- Globalization's effect on infrastructure revitalization in the U.S.
- The dependency of economic wealth on the variety, reliability, and service of physical infrastructure systems
- Cost and governmental issues
- The necessity to be actively involved in underdeveloped countries

## Environment

How will globalization impact the civil engineer's approaches toward, and abilities to deal with, environmental issues in the 21st century? Possible issues include:

- The international scale of extreme and long-term environmental events, such as disasters, global climate change, and their impacts on the natural, built, and social environments
- Meeting a world health standard
- Developing acceptable international standards for both large and developing countries

## Computer Tools and the Internet

How will various computer tools and the Internet impact the civil engineering profession in carrying out international collaboration and project management in the 21st century? Possible issues include:

- The rapid advances made in computation and Internet usability and their impact on project management
- International collaboration and data transfer using the Internet
- Outsourcing and doing business remotely using the Internet and other computer tools

## Definitions of Globalization

There are many definitions of globalization. However, those relevant to the civil engineering profession are provided below:

- Development of extensive worldwide patterns of economic relationships between nations. www.investorwiz.com/glossary.htm
- Globalization refers in general to the worldwide integration of humanity and the compression of both the temporal and spatial dimensions of planet wide human interaction. www2.truman.edu/~marc/resources/terms.html
- The increasing worldwide integration of markets for goods, services, and capital that attracted special attention in the late 1990s. Also used to encompass a variety of other changes that were perceived to occur at about the same time, such as an increased role for large corporations in the world economy and increased intervention into domestic

policies and affairs by international institutions such as the IMF, WTO, and World Bank. www.personal.umich.edu/~alandear/glossary/g.html

- A set of processes leading to the integration of economic, cultural, political, and social systems across geographical boundaries.
  www.hsewebdepot.org/imstool/GEMI.nsf/WEBDocs/Glossary

- The process of developing, manufacturing, and marketing software products that are intended for worldwide distribution. This term combines two aspects of the work: internationalization (enabling the product to be used without language or culture barriers) and localization (translating and enabling the product for a specific locale).
  www.cit.gu.edu.au/~davidt/cit3611/glossary.htm

- The generalized expansion of international economic activity, which includes increased international trade, growth of international investment (foreign investment) and international migration, and increased creation of technology among countries. Globalization is the increasing worldwide integration of markets for goods, services, labor, and capital.
  minneapolisfed.org/econed/essay/topics/glossary05.cfm

- The movement toward markets or policies that transcend national borders.
  www.wcit.org/tradeis/glossary.htm

- Tendency of integration of national capital markets.
  www.equanto.com/glossary/g.html

- In the translation/localization business marketplace, it refers to the whole problem of making any product or service global, with simultaneous release in all markets. Web site globalization means more than making one Web site respond to the different language and regional requirements of the browser.
  www.openinternetlexicon.com/Glossary/GlobalGlossary.html

- A process of creating a product or service that will be successful in many countries without modification.
  www.bena.com/ewinters/Glossary.html

- Trend away from distinct national economic units and toward one vast global market. enbv.narod.ru/text/Econom/ib/str/261.html

- Used for transnational influences on culture, economics, politics, et cetera—especially illustrating global patterns or trends.
  lib.ucr.edu/depts/acquisitions/YBP%20NSP%20GLOSSARY%20EXTERNAL%20revised6-02.php

- In the modern global economy no country can sustain itself as a closed economy.
  www.sasked.gov.sk.ca/curr_content/entre30/helppages/glossary/glossary.html

- The increasing economic, cultural, demographic, political, and environmental interdependence of different places around the world.
  hhhknights.com/geo/4/agterms.htm

- A relatively new word that is commonly used to describe the ongoing, multidimensional process of worldwide change. It describes the idea that the world is becoming a single global market. It describes the idea that time and space have been shrunk as a result of modern telecommunications technologies which allow almost instantaneous communication between people almost anywhere on the planet.
  www.takebackwisconsin.com/Documents/Glossary.htm

- The increasing integration of world markets for goods, services, and capital.

It has also been defined as a process by which nationality becomes increasingly irrelevant in global production and consumption.
www.agtrade.org/glossary_search.cfm

- The integration of markets on a worldwide scale could eventually mean worldwide standards or practices for product quality, pricing, service, and design.
www.ucs.mun.ca/~rsexty/business1000/glossary/G.htm
- It refers to international exchange or sharing of labor force, production, ideas, knowledge, products, and services across borders.
www.kwymca.org/nccq/glossary.htm
- The intensification of worldwide social relations, which, through economic, technological and political forces, link distant localities in such a way that distant events and powers penetrate local events.
www.anthro.wayne.edu/ant2100/GlossaryCultAnt.htm
- The process of making something worldwide in scope or application.
schools.cbe.ab.ca/logistics/g.html
- Refers to the widening, deepening, and speeding up of worldwide interconnectedness in all aspects of contemporary life. (All aspects, including its nature, causes, and effects are hotly disputed, with strange bedfellows on all sides.)
www.ripon.edu/academics/global/CONCEPTS.HTML
- Is a term used to refer to the expansion of economies beyond national borders—in particular, the expansion of production by a firm to many countries around the world—that is, globalization of production, or the "global assembly line." This has given transnational corporations power beyond nation-states, and has weakened any nation's ability to control corporate practices and flows of capital, set regulations, control balances of trade and exchange rates, or manage domestic economic policy.
colours.mahost.org/faq/definitions.html
- A contested term relating to the transformation of spatial relations that involves a change in the relationship between space, economy, and society.
media.pearsoncmg.com/intl/ema/uk/0131217666/student/0131217666_glo.html
- Growth to a global or worldwide scale; "the globalization of the communication industry."
wordnet.princeton.edu/perl/webwn
- Globalization (or globalisation) is a term used to describe the changes in societies and the world economy that are the result of dramatically increased trade and cultural exchange. In specifically economic contexts, it is often understood to refer almost exclusively to the effects of trade, particularly trade liberalization or free trade.
en.wikipedia.org/wiki/Globalization

## Sources


Arciszewski, T. 2006. "Civil Engineering Crisis," *Journal of Leadership and Management in Engineering—ASCE*, pp. 26–30, January.

ASCE Body of Knowledge Committee. 2004. *Civil Engineering Body of Knowledge for the 21st Century: Preparing the Civil Engineer for the Future*, Reston, VA, January. (http://www.asce.org/raisethebar)

http://www.google.com/search?hl=en&defl=en&q=define:Globalization&sa=

X&oi=glossary_definition&ct=title (accessed, April 2, 2007)

Jha, M.K. and D. Lynch. 2007. "Role of Globalization and Sustainable Engineering Practice in the Future Civil Engineering Education," in Sustainable Development and Planning, III, A Kungolos, C. A. Brebbia, E. Beriatos (eds), Vol. 2, 641–650.

National Academy of Engineering. 2006. *Rising Above the Gathering Storm: Energizing and Employing America for a Brighter Economic Future,* National Academies Press, Washington, D.C.

# APPENDIX N

# Public Policy

Individuals entering the professional practice of civil engineering must recognize that civil engineering activities are not conducted in isolation from the general public. Civil engineers need to understand the engineering/public policy interface and how decision makers in government utilize technical, scientific, and economic information when planning, designing, or evaluating civil engineering projects. Continuous integration of civil engineers into the public policy arena is critical to the well-being of society at large.

Public policy is the articulation of the nation's, state's, or municipality's goals and values. Civil engineers are most involved in public policy regarding both the physically built environment and the preserved natural environment. Essential public policy fundamentals include the political process, laws/regulations, funding mechanisms, public education, government/business interaction, and the public service responsibility of professionals. These issues heavily influence many civil engineering decisions.

In comparison to many other engineering disciplines, the practice of civil engineering is unique with respect to public policy. Civil engineering systems have a much broader societal context requiring a firm understanding of public policy development. The core economics of the profession are shaped by public policy debates and decisions. A substantial portion of financial support for civil infrastructure and preservation of the natural environment comes from public funds. Additionally, the use of the built environment is generally available for public consumption and protected natural areas are typically accessible for passive public use. Practitioners need to be active participants in the process across all aspects of civil engineering.

Public discourse and debate are key components associated with the funding, development, maintenance, and rehabilitation of our country's infrastructure. At the outset of one's career, civil engineers must understand how their work relates to the nation's public policy. Professional engineers possess a vast quantity of knowledge and experience that is extremely valuable in the development of public policy. As they continue to grow during their career, the civil engineers become increasingly involved in the development of public policy. Future civil engineers must attain quality technical knowledge and skills, and also have the ability to assist society in understanding the complex nature of building a safe and sustainable physical environment that supports the needs of the community.

Literature on public policy cites many and varied definitions of public policy. Thomas Dye provides a very a concise definition: "whatever governments choose to do or not to do."[a] Because civil engineering is often referred to as a people-serving profession, and the people are the public, civil engineers are inherently and increasingly more fully engaged in the public policy process. Whether publicly or privately held, the civil infrastructure and the natural environment are engineered works that directly affect the daily lives of people. Civil engineers naturally practice their profession by following standards, specifications, and related guidelines as set forth in public policy documents. Accordingly, engineering students should be exposed to the overall process of formulating public policy.

Civil engineers interact regularly with public officials, decision makers, and agencies. Conflicts can arise in what constitutes "the public interest" and disagreements often entail what makes public policy public.[b] Thomas Birkland offers several key considerations regarding the study and practice of public policy:[b]

- The practical reasons for studying policy are political reasons.
- It is important to understand the process that leads to decisions to make policy statements.
- Those who understand public policy processes are best able to meet their respective policy goals.

The range of public policy issues, processes, and implementation begins with an awareness and understanding of public policy procedures, progresses via systematic evaluation of potential outcomes from public policy decisions, and generally culminates in policy designs and tools that guide future public interest decisions.

Diane Rover argues that policy issues can take the form of policy making created by engineers and policy making to create engineers.[c] Within the civil engineering BOK context, both are relevant. Rover provides a summary review of the National Academy of Engineering report *The Engineer of 2020: Visions of Engineering in the New Century* (2004). Several of the cited NAE needs and goals align naturally with the future preparation of civil engineers:

- By 2020, engineers will assume leadership positions that can have positive influences in public policy issues related to government and industry.
- Future engineers will need to be engage more effectively in policy issues.
- The convergence between engineering and public policy will increase as technology becomes more permanently engrained into society.
- Likewise, engineers will need to understand the policy by-products of new technologies and be public servants who recognize the implications of related policy decisions.

*The Engineer of 2020*[d] clearly states the concerns that while the United States has the "best physical infrastructure in the developed world.... these infrastructures are in serious decline ... and are among the top concerns for public officials and citizens alike." Because the built infrastructure is the domain of civil engineers, the civil engineer of the future must be a full participant in the policy decisions justifying required investments to address the aging civil infrastructure. The NAE enumerates key engineering roles that are anticipated regarding public policy. The following items are directly associated with, affected by, and/or mitigated via the civil engineering profession:

- Environment

- Energy
- Health
- Education
- Water supply and quality

The social context of engineering practice as expressed in *Educating the Engineer of 2020*[e] acknowledges that although the future is uncertain, "engineering will not operate in a vacuum separate from society in 2020." This NAE report provides two other relevant insights:

The professional context for engineers in the future necessitates excellent communication skills when addressing both technical and public audiences.

The convergence between engineering and public policy is reinforced, as is the requirement to responsibly articulate the policy issues affecting the general public.

Civil engineers will most certainly be at a natural nexus to meaningfully participate in the envisioned public debates. The debates can include working directly with Congress and other national public, private, and professional organizations. Beginning in the early 1990's, various groups were calling upon scientists and engineers to build stronger relationships with their respective senators and representatives.[f] In his practical guide "Working with Congress," William Wells presents several justifications for engaging in policy making by scientists and engineers:[f]

- It is important not to leave science and technology policy issues in the hands of other interests groups.
- It should be made clear to decision makers that scientific ideas (and engineering matters) are based upon generally accepted data and analysis.
- Working with Congress does serve the public and national interest, the professions, and associated institutions' self-interests.
- Most notably, to ignore Congress would abdicate one's responsibility to the science and engineering communities.

The Eno Transportation Foundation offers specific guidance and insights on the policy development process and outcomes. Succinctly, stakeholders need to have a clear understanding of how government develops and implements policies. It is likewise important to know how one can participate in shaping public policy.[g] Since our American system allows each stakeholder group to have a voice in establishing public policy, the civil engineer, and the civil engineering profession, should actively participate in the process. Two most recent and noteworthy examples of active participation by scientists and engineering in setting public policy are the NAE report *Engineering Research and America's Future*,[h] and The National Academies report titled: *Rising Above the Gathering Storm*.[i]

## Cited Sources

a) Dye, T. R. 1992. *Understanding Public Policy.* 7th ed. Prentice Hall. Englewood Cliffs, NY.

b) Birkland, T. A. 2005. *An Introduction to the Policy Process: Theories, Concepts, and Models of Public Policy Making.* 2nd ed. M.E. Sharpe, London, England.

c) Rover, D. 2006. "Policy Making and Engineers," *Journal of Engineering Education*, pp.93–95, January.

d) National Academy of Engineering. 2004. *The Engineer of 2020: Visions of Engineering in the New Century.* The National Academies of Sciences, Washington, D.C. (http://www.nae.edu)

e) National Academy of Engineering. 2005. *Educating the Engineer of 2020: Adapting Engineering Education to the New Century,* National Academy of Sciences, Washington, DC. (http://www.nae.edu)

f) Wells, Jr., W. G. 1992. Working with Congress. 2nd ed., American Association for the Advancement of Science, Washington, D.C.

g) Eno Transportation Foundation. 2005. *National Transportation Organizations: Their Roles in the Policy Development and Implementation Process.* The Eno Transportation Foundation, Washington, D.C.

h) National Academy of Engineering. 2005. *Engineering Research and America's Future: Meeting the Challenges of a Global Economy.* The National Academies Press, Washington, D.C.

i) National Academy of Engineering. 2006. *Rising Above the Gathering Storm: Energizing and Employing America for a Brighter Economic Future,* National Academies Press, Washington, D.C.

# APPENDIX O
# Attitudes

## Findings of the First Body of Knowledge Committee

The first BOK report[a] defined the ASCE BOK as the knowledge, skills, and attitudes necessary for an individual to enter the professional practice of civil engineering. Several reasons were listed in the report as to why "attitudes" were included in the definition. Some of those arguments are repeated here as is information defending the proposition of including "attitudes" as its own outcome in this second edition of the ASCE BOK.

Attitudes refer to the "ways in which one thinks and feels in response to a fact or situation."[b] At the professional level, one's attitudes will affect how knowledge and skills are applied to the solution of a civil engineering problem. The first BOK committee provided three reasons for including attitudes in the definition of a BOK:[c]

- A wealth of study and professional opinion points to the importance of attitude in individual and group achievement.
- Teaching of attitudes is an integral part of educational practice.
- Attitudes are an integral part of the BOKs of other professions and specialties such as architecture, accounting, and law.

The first BOK committee noted that "knowledge and skills are more comfortably and frequently discussed by civil engineers and probably many other professionals. This tendency is explained, in part, by the objectivity and specificity of knowledge and skills in contrast to the subjectivity and ambiguity of attitudes."[d] Attitudes are more difficult to assess than knowledge and skills, yet attitudes affect behavior, which certainly can be measured.

An exhaustive list of appropriate attitudes would be difficult to compile. In the present case, the significant attitudes are those that support the effective practice of civil engineering. A partial list of those attitudes might include commitment, confidence, consideration of others, curiosity, fairness, high expectations, honesty, integrity, intuition, good judgment, optimism, persistence, positiveness, respect, self-esteem, sensitivity, thoughtfulness, thoroughness and tolerance.[e]

The first BOK report addressed the question, can attitudes be taught? The report notes that attitudes certainly can be taught, but the essential question is whether students or pre-professional engineers can learn appropriate attitudes.[f] The subsequent challenge is to encourage the professional community to adopt, practice, and assess those attitudes that are supportive of the effective practice of civil engineering.[g]

In conclusion, the first BOK committee writes:

> *"Despite the complications of subjectivity and ambiguity, the BOK Committee members are convinced that attitudes must join knowledge and skills as one of the three essential components of the "what" dimension of the civil engineering BOK. The manner in which a civil engineer views and approaches his or her work is very likely to determine how effectively he or she uses hard-earned knowledge and skills."*[h]

A more detailed version of the material presented here can be found in the paper by Hoadley.[i]

## The Importance of Attitudes in the Engineering Profession and Beyond

The authors of the first BOK report cited several articles that noted the importance of attitudes to the engineering profession. A few more are considered here. For example, Elms noted that "besides having good technical training, a professional engineer has something more, which distinguishes him [or her] from a technician. The extra quality is a set of attitudes, some of which—holism, realism, and flexibility—can be encouraged by university teaching."[j] Stouffer wrote that a particular set of attitudes is important in effective engineering management.[k] With regard to management efficiency, Kahn notes the inculcation of appropriate attitudes is necessary in the manufacturing engineering profession.[l] In a survey of job advertisements for engineering professionals, Henshaw found that that employers wanted applicants who possess good communication skills, work well on teams, possess the ability to relate to people, and hold positive attitudes.[m]

Other professions recognize the importance of constructive attitudes in the successful completion of a task. Janke, when addressing educators in pharmaceutical schools, emphasized the importance of attitudes in the effectiveness of teaching and learning.[n] Morgan states that the competencies for software professionals "have been defined as a set of observable performance dimensions, knowledge, attitudes, [and] behavior, as well as collective team, process, and organizational capabilities that are linked to high performance."[o] There seems to be much support for the idea that attitudes are important for engineering and other professionals.

## Attitudes or Abilities?

Some on the second BOK committee suggested that the civil engineering BOK should include "abilities" with "knowledge" and "skills" rather than "attitudes." Indeed the term "abilities" is used rather than "attitudes" in several job descriptions developed by the computer science profession,[p] the California[q] and Oklahoma[r] state licensing boards, the Office of Aviation Safety,[s] and the U.S. Office of Personnel Management.[t]

Necessary knowledge, skills, and "abilities" are indeed listed as prerequisites for professional practice in some cases; however, in many cases necessary "attitudes" are also specified including examples in the engineering community,[u] the academic community,[v,w] the human resource profession,[x] health care profession[y,z] and others.[aa,bb,cc]

The term "attitudes" is used in many professional circles as a part of their respective BOKs or equivalent. ASCE is not outside of professional practice when requiring certain "attitudes" within its BOK.

## Assessing Attitudes

One of the concerns when including "attitudes" in the BOK is the difficulty of assessing them in a meaningful way. Knowledge and skills can be objectively measured but attitudes are far more subjective. Any given measurement of one's attitude is plagued by a host of uncertainties. For example, the subject, by his or her own actions and words, may hide his or her true attitude regarding a particular topic for a host of reasons. An observer may distort an accurate measure of another's attitudes. The development of an appropriate assessment scheme will require much time and effort so a thorough discussion of the topic will not be attempted here. Even so, a few comments may prove helpful.

Perhaps the simplest way to measure the attitude of a licensure applicant would be through an assessment by a supervising professional engineer. Many licensing agencies already require some measurement of such subjective qualities as character and integrity. The state of California requires four references who must rate an engineering applicant's technical competency, judgment, and integrity among other characteristics.[dd] In North Carolina references must rate an applicant's integrity and ethical behavior,[ee] and in Oklahoma applicants must be technically competent and of good character as attested to by at least five references.[ff]

Thurstone was one of the pioneers in the measurement of attitudes. His methods for measuring attitudes use a simple agree/disagree scale. This approach involves two main stages. The first is to develop a large number of attitude statements regarding a topic. Subjects are then asked to rate how they agree or disagree with the attitude statements.[gg] This method requires multiple attitude statements before an accurate measure of one's attitude can be obtained.

The Semantic-Differential method of measuring attitudes devised by Osgood consists of a topic and a set of bipolar scales—for example, exciting to dull.[hh] The subject has to indicate the direction and intensity of an attitude towards a given topic. The design of the statement and the scales are important in the accurate assessment of attitude.

The assessment of attitudes has been a long study within several professions. For example, the impact of a teacher's and a student's attitudes on learning has long been a study in the education profession.[ii] Even though it may be difficult for the engineering profession to develop an efficient assessment tool, attempts have been made in other professions; therefore, it is entirely appropriate for the engineering community to pursue the same.

## Concluding Remarks

A BOK for any profession certainly includes knowledge and skills. Because attitudes affect how the knowledge and skills are used, attitudes should be included in the BOK. Attitudes are specified in the equivalent BOKs of other professions and the assessment of attitudes has long been a study. The inclusion of this outcome in the new edition of the BOK certainly "raises the bar" for the profession.

## Cited Sources


a) ASCE Body of Knowledge Committee. 2004. *Civil Engineering Body of Knowledge for the 21st Century: Preparing the Civil Engineer for the Future*, Reston, VA, January. (http://www.asce.org/raisethebar).

b) ibid

c) ibid

d) ibid

e) ibid

f) ibid

g) ibid

h) ibid

i) Hoadley, P. W. 2007. "The BOK and Attitudes Assessment," Proceedings of the ASEE Annual Conference, June.

j) Elms, D. G., "Steps Beyond Technique: Education for Professional Attitude," *Civil Engineering Systems,* 2(1), 55–59, 1985.

k) Stouffer, W. B., J. S. Russell, and M. G. Oliva. 2004. "Making the Strange Familiar: Creativity and the Future of Engineering Education," Proceedings of the ASEE Annual Conference, American Society for Engineering, Washington, DC, 9315–9327.

l) Khan, H. 1996. "Correlates of Engineering and Management Effectiveness: Design of a Strategic University Curriculum for Corporate Engineering Executive Development (SUCCEED) program." Proceedings of the 26th Annual Conference on Frontiers in Education, Part 2 (of 3), IEEE, Nov 6–9 1996, Piscataway, NJ 886–890.

m) Henshaw, R. 1991. "Survey of Professional Engineering Job Advertisements," International Mechanical Engineering Congress and Exhibition.

n) Janke, K. K. Accreditation Council for Pharmacy Education webpage, http://www.acpe-accredit.org/scholar05/Breeze/AssessingAttitudes/index.html.

o) Morgan, J. N. 2005. "Why the Software Industry Needs a Good Ghostbuster," *Communications of the ACM,* v48(8).

p) Bailey, J. L. and G. Stefaniak. 2001. "Industry Perceptions of the Knowledge, Skills, and Abilities Needed by Computer Programmers," Proceedings of the ACM SIGCPR Conference on Computer Personnel Research.

q) California Board for Professional Engineers and Land Surveyors. 2007. "Professional Engineers Act," January 1. http://www.dca.ca.gov/pels/e_plppe.pdf.

r) Oklahoma State Board of Licensure for Professional Engineers & Land Surveyors Web page, http://www.pels.state.ok.us/.

s) Office of Aviation Safety of the National Transportation Safety Board, http://www.ntsb.gov/vacancies/descriptions/AeroEngPowerplants.doc.

t) U.S. Office of Personnel Management, http://www.opm.gov/qualifications/SEC-IV/B/GS0800/0800.htm.

u) Marshall and Marshall. 2005. "Facilitating the Development of Student's Personal Ethics in Cultivating Professional Ethics in Engineering Classrooms," Proceedings of the ASEE Annual Conference and Exposition, June, 6323–6330.

v) Quádernas, César. 2000. "Improving Academic Performance Through Typifying Electronics Engineers," Proceedings,

Frontiers in Education Conference, IEEE.

w) Fullen, M. G., B. Joyce, and B. Showers. 2002. *Student Achievement through Staff Development*, 3rd Edition, Association for Supervision and Curriculum Development, Alexandria, VA.

x) Markman, G. D. 2003. "Person-Entrepreneurship Fit: Why Some People Are More Successful as Entrepreneurs than Others," *Human Resource Management Review*, 13(2).

y) Hodges, B., C. Inch, and I. Silver. 2001. "Improving the Psychiatric Knowledge, Skills and Attitudes of Primary Care Physicians," *American Journal of Psychiatry*, 158, October.

z) Lennox, N., and J. Diggens. 1999. "Knowledge, Skills and Attitudes: Medical Schools' Coverage of an Ideal Curriculum on Intellectual Disability," *Journal of Intellectual & Developmental Disability*, 24(4).

aa) Freeman, M. 2004. "SLT Talk and Practice Knowledge: A Response to Ferguson and Armstrong," *International Journal of Language & Communication Disorders*, 39(4).

bb) Kääriäinen, J., P. Sillanaukee, P. Poutanen, and K. Seppä. 2001. "Opinions on Alcohol-Related Issues Among Professionals in Primary, Occupational, and Specialized Health Care," *Alcohol and Alcoholism*, 36(2).

cc) Rodolfa et al. 2005. "A Cube Model for Competency Development: Implications for Psychology Educators and Regulators," *Professional Psychology: Research and Practice*, 36(4).

dd) California Board for Professional Engineers and Land Surveyors, PE application, http://www.dca.ca.gov/pels/a_appinstpe.htm.

ee) North Carolina Board of Examiners For Engineers and Surveyors Web site, http://www.ncbels.org/reg-pe.htm

ff) Oklahoma State Board of Licensure for Professional Engineers & Land Surveyors Web page, http://www.pels.state.ok.us/

gg) Roberts, J. S., J. E. Laughlin, and D. H. Wedell. 1997. "Comparative Validity of the Likert and Thurstone Approaches to Attitude Measurement," *ETS Report*.

hh) Osgood, C. E., G. J. Suci, and P. H. Tannenbaum. 1957. "The Measurement of Meaning," University of Illinois Press, Urbana.

ii) Karavas-Doukas, E. 1996. "Using Attitude Scales to Investigate Teachers' Attitudes to the Communicative Approach, *ELT Journal*, 50(3).

# APPENDIX P

# Notes

1. ASCE Task Committee to Plan a Summit on the Future of the Civil Engineering Profession. 2007. *The Vision for Civil Engineering in 2025,* Reston, VA, ASCE. (A PDF version is available, at no cost, from http://www.asce.org/Vision2025.pdf)
2. ASCE Policy Statement 465 as adopted by the ASCE Board of Direction on April 24, 2007. See the "Issue" section. (http://www.asce.org/raisethebar)
3. ASCE Body of Knowledge Committee. 2004. *Civil Engineering Body of Knowledge for the 21st Century: Preparing the Civil Engineer for the Future,* Reston, VA, January. (http://www.asce.org/raisethebar).
4. National Academy of Engineering. 2004. *The Engineer of 2020: Visions of Engineering in the New Century,* The National Academies of Sciences, Washington, DC. (http://www.nae.edu)
5. National Academy of Engineering. 2005. *Educating the Engineer of 2020: Adapting Engineering Education to the New Century,* National Academies of Sciences, Washington, DC. (http://www.nae.edu)
6. This is the ASCE definition of civil engineering, as adopted in 1961 by the ASCE Board of Direction and published in the ASCE *Official Register.*
7. In late 2007, ASCE formed the Task Committee to Achieve the Vision for 2025.
8. For the current complete ASCE Policy Statement 465, see http://www.asce.org/raisethebar.
9. ASCE Levels of Achievement Subcommittee. 2005. *Levels of Achievement Applicable to the Body of Knowledge Required for Entry Into the Practice of Civil Engineering at the Professional Level,* Reston, VA, September. (http://www.asce.org/raisethebar)
10. Bloom. B. S., Englehart, M. D., Furst. E. J., Hill, W. H., and Krathwohl, D. 1956. *Taxonomy of Educational Objectives, the Classification of Educational Goals, Handbook I: Cognitive Domain,* David McKay, New York, NY.
11. Rohwer, W. D., Jr. and K. Sloane. 1994. "Psychological Perspectives." In L. W. Anderson and Sosniak L.A., "Bloom's Taxonomy: A Forty-Year Retrospective," *Ninety-third Yearbook of the National Society for the Study of Education,* University of Chicago Press, pp. 41–63, Chicago, IL.
12. ABET. 2007. *Criteria for Accrediting Engineering Programs: Effective for Evaluations During the 2007–2008 Accreditation Cycle,* ABET, Inc., Baltimore, MD. (http://www.abet.org).

13. Pre-licensure experience consists primarily of work during the pre-licensure period but for some engineers will also include other relevant experience such as active involvement in professional societies and community affairs.

14. According to Merriam-Webster, rubric is defined as "an authoritative rule" and "something under which a thing is classed" (http://www.m-w.com/dictionary/rubric.). Cambridge Learner's Dictionary defines rubric as "a set of instructions or an explanation" (http://dictionary.cambridge.org/define.asp?dict=L&key=HW*58100112)

15. ASCE Body of Knowledge Fulfillment and Validation Committee. 2005. *Fulfillment and Validation of the Civil Engineering Body of Knowledge,* Reston, VA, April. (http://www.asce.org/pdf/FVReportFinal.pdf)

16. ASCE Experience Committee. 2007. *Final Report,* Reston, VA, July (http://www.asce.org/raisethebar).

17. ASCE Task Committee on the Academic Prerequisites for Professional Practice (TCAP[3]). 2003. "ASCE's Raise the Bar Initiative: Master Plan for Implementation," Session No. 2315, *Proceedings of the American Society for Engineering Education Annual Conference and Exposition,* June 22–25, Nashville, TN. The Master Plan was first developed by TCAP[3] in 2002 and the cited paper is one of the first times it was published. The Master Plan has since been refined with the current version being Figure 4 in this BOK2 report.

18. ASCE Curriculum Committee. 2007. *Development of Civil Engineering Curricula Supporting the Body of Knowledge,* Reston, VA, December. (http://www.asce.org/raisethebar).

19. See http://www.asce.org/files/pdf/professional/BLPCALGCV35b.pdf.

20. See http://www.asce.org/files/pdf/professional/ASCECommentaryv3.309232006.pdf.

21. Engineers are licensed in 50 states plus the District of Columbia and four U.S. territories (Guam, Puerto Rico, Northern Mariana Islands, and the Virgin Islands) for a total of 55 licensing jurisdictions. Illinois has a separate board for structural engineering. Therefore, there are 56 boards that license engineers. For an historical account of U.S. engineering licensure, see McGuirt, D. 2007, "The Professional Engineering Century," *PE,* June, pp. 24–29 and for thoughts on the future of licensure, see Nelson, J. D. and B. E. Price, 2007, "The Future of Professional Engineering Licensure, *PE,* June, pp. 30–34.

22. ASCE. 1995. *Summary Report—1995 Civil Engineering Education Conference (CEEC '95).*

23. Adelman, C. 1998. *Women and Men of the Engineering Path: A Model for Analysis of Undergraduate Careers,* U.S. Department of Education, Washington, D.C.

24. Astin, A. W., and H. S. Astin. 1993. *Undergraduate Science Education: The Impact of Different College Environments on the Educational Pipeline of the Sciences.* Los Angeles Higher Education Resource Institute, UCLA.

25. Boyer, E. L. 1990. *Scholarship Reconsidered: Priorities of The Professoriate, A Special Report.* The Carnegie Foundation for the Advancement of Teaching.

26. Glassick, C. E., M. T. Huber, and G. I. Maeroff. 1997. *Scholarship Assessed: Evaluation of the Professoriate,* Special Report of the Carnegie Foundation for the Advancement of Teaching, Jossey-Bass Inc., San Francisco, CA.

27. Lowman, J. 1995. *Mastering The Techniques of Teaching,* Jossey-Bass, Inc., San Francisco, CA.

28. Conley, C., Ressler, S., Lenox, T., and Samples, J. 2000. "Teaching Teachers to Teach Engineering," *Journal of Engineering Education*, ASEE, January.

29. Dennis, N. 2001. "ExCEEd Teaching Workshop: Taking It on the Road," *Proceedings of the American Society for Engineering Education*, Albuquerque, NM.

30. Douglas, E. 2001. "A Comprehensive Approach to Classroom Teaching: Does It Work?" *Proceedings of the American Society for Engineering Education*, Albuquerque, NM.

31. Estes, A., and S. Ressler. 2001. "ExCEEd Teaching Workshop: Fulfilling a Critical Need," *Proceedings of the American Society for Engineering Education*, Albuquerque, NM.

32. Knapp, K. 2000. "Learning to Teach Engineers: The Applicability and Compatibility of One Approach," *Proceedings of the American Society for Engineering Education*, St. Louis, MO.

33. Welch, R., Baldwin, J., Bentler, D., Clarke, D., Gross, S., and Hitt, J. 2001. "The ExCEEd Teaching Workshop: Participants' Perspective and Assessment," *Proceedings of the American Society for Engineering Education*, Albuquerque, NM.

34. Welch, R., Baldwin, J., Bentler, D., Clarke, D., Gross, S., and Hitt, J. 2001. "The ExCEEd Teaching Workshop: Hints to Successful Teaching," *Proceedings of the American Society for Engineering Education*, Albuquerque, NM.

35. Examples of books that may help the student successfully complete his or her studies and proactively move into employment are:

    Berson, B. R. and D. E. Benner. 2007. *Career Success in Engineering: A Guide for Students and New Professionals,* Kaplan AEC Education, Chicago, IL.

    Goldberg, D. E. 1995. *Life Skills and Leadership for Engineers,* McGraw Hill, New York, NY.

    Roadstrum, W. H. 1998. *Being Successful As An Engineer,* Engineering Press, Austin, TX.

    Walesh, S. G. 2000. *Engineering Your Future: The Nontechnical Side of Professional Practice in Engineering and Other Technical Fields,* ASCE Press, Reston, VA.

    Walesh, S. G. 2004. *Managing and Leading: 52 Lessons Learned for Engineers,* ASCE Press, Reston, VA.

36. Mandino, O. 168. *The Greatest Salesman in the World,* Bantam Books, New York, NY.

37. Neufeldt, V., Editor in Chief. 1986. *Webster's New World Dictionary of American English,* Third College Edition, Prentice Hall, New York, NY.

38. Foundation for Critical Thinking, "The Critical Thinking Community," (http://www.criticalthinking.org/aboutCT/definingCT.shtml).

39. Lipman, M. 1988. "Critical Thinking: What Can It Be?," *Education Leadership,* Vol. 46, No. 1., pp. 38–43.

40. Van Joolingen, W. 1999. "Cognitive Tools for Discovery Learning," *International Journal of Artificial Intelligence in Education,* Vol. 10, pp. 385–397.

41. This is the ASCE definition of sustainable development, adapted to describe the ability of engineering activity to meet its service goals.

42. This is the ASCE definition of sustainable development, as adopted in 1996 by the ASCE Board of Direction and recognized since then in the ASCE Code of Ethics (http://www.asce.org/inside/codeofethics.cfm). It is the root of other sustainability definitions that appear in this report's Appendix B, Glossary.

43. *ASCE Code of Ethics,* Fundamental Canon 1: "Engineers shall hold paramount the safety, health and welfare of the public and shall strive to comply with the principles of sustainable development in the performance of their professional duties." The ASCE definition of Sustainable Development (November 1996), and used here, is recorded therein. https://www.asce.org/inside/codeofethics.cfm

44. This is the ASCE definition of sustainable development adapted to describe the engineering challenge.

45. ASCE Task Committee on the First Professional Degree. 2001. *Engineering the Future of Civil Engineering,* Reston, VA, October 9. (http://www.asce.org/raisethebar.)

46. *Merriam-Webster Dictionary,* online at: http://www.m-w.com/cgi-bin/dictionary

47. Parcover, J. A. and R. H. McCuen. 1995. "Discovery Approach to Teaching Engineering Design," *Journal of Professional Issues in Engineering Education—ASCE,* Vol. 121, No. 4, pp. 236–241.

48. *ASCE Policy Statement 418, The Role of the Civil Engineer in Sustainable Development,* adopted by the ASCE Board of Direction, October 19 2004.

49. *ASCE Policy Statement 517, Millennium Development Goals,* adopted by the ASCE Board of Direction, July 22, 2006.

50. Lynch D., W. Kelly, M. K. Jha, and R. Harichandran. "Implementing Sustainably in the Engineering Curriculum: Realizing the ASCE Body of Knowledge," proceedings of the ASEE Annual Conference, Honolulu, HI, June 2007.

51. ASCE Task Committee on Sustainability Criteria and UNESCO/IHP IV Project M-4.3, D.P. Loucks et al. 1998. *Sustainability Criteria for Water Resource Systems,* Reston, VA.

52. ASCE Committee on Sustainability. 2004. *Sustainable Engineering Practice: An Introduction,* Reston, VA. Principles collected from several sources are summarized at p.96 ff.

53. "Physical" here refers to the domain of the physical sciences, as distinct from the social sciences. For example, included are physics, chemistry, biology, and the earth sciences.

54. Vest, C. M. 2006. "Educating Engineers for 2020 and Beyond." *The Bridge,* National Academy of Engineering, Summer, pp 39–44.

55. Petroski, H. 2001. "The Importance of Engineering History," *International Engineering History and Heritage.* ASCE, Reston, VA.

56. *Webster's New World Dictionary.* 1988. Third College Edition.